D1625085

In-Memory Data Management

Hasso Plattner · Alexander Zeier

In-Memory Data Management

Technology and Applications

Second Edition

Hasso Plattner
Hasso Plattner Institute
Potsdam, Brandenburg
Germany

Alexander Zeier
Massachusetts Institute of Technology
Cambridge, MA
USA

ISBN 978-3-642-29574-4   ISBN 978-3-642-29575-1 (eBook)
DOI 10.1007/978-3-642-29575-1
Springer Heidelberg New York Dordrecht London

Library of Congress Control Number: 2012935682

© Springer-Verlag Berlin Heidelberg 2011, 2012
This work is subject to copyright. All rights are reserved by the Publisher, whether the whole or part of the material is concerned, specifically the rights of translation, reprinting, reuse of illustrations, recitation, broadcasting, reproduction on microfilms or in any other physical way, and transmission or information storage and retrieval, electronic adaptation, computer software, or by similar or dissimilar methodology now known or hereafter developed. Exempted from this legal reservation are brief excerpts in connection with reviews or scholarly analysis or material supplied specifically for the purpose of being entered and executed on a computer system, for exclusive use by the purchaser of the work. Duplication of this publication or parts thereof is permitted only under the provisions of the Copyright Law of the Publisher's location, in its current version, and permission for use must always be obtained from Springer. Permissions for use may be obtained through RightsLink at the Copyright Clearance Center. Violations are liable to prosecution under the respective Copyright Law.
The use of general descriptive names, registered names, trademarks, service marks, etc. in this publication does not imply, even in the absence of a specific statement, that such names are exempt from the relevant protective laws and regulations and therefore free for general use.
While the advice and information in this book are believed to be true and accurate at the date of publication, neither the authors nor the editors nor the publisher can accept any legal responsibility for any errors or omissions that may be made. The publisher makes no warranty, express or implied, with respect to the material contained herein.

Printed on acid-free paper

Springer is part of Springer Science+Business Media (www.springer.com)

In Praise of "In-Memory Data Management"

Academia

Prof. David Simchi-Levi (Massachusetts Institute of Technology, Cambridge, USA)

The book "In-Memory Data Management—Technology and Applications" by Hasso Plattner and Alexander Zeier describes a revolutionary database technology and many implementation examples for business intelligence and operations. Of particular interest to me are the great opportunities opening up in supply chain management, where the need to balance the speed of planning algorithms with data granularity has been a long time obstacle to performance and usability.

Prof. Karl Max Einhäupl (CEO, Charité—Universitätsmedizin Berlin, Germany)

Personalized health care requires the combination of distributed medical data. In-memory technology enables us to analyze millions of patient records within a second and allows us to devise therapies even closer to the patients' individual needs than ever before.

Prof. Christoph Meinel (Hasso Plattner Institute (HPI), Potsdam, Germany)

I'm proud that HPI and the cooperation between HPI and SAP has provided such an inspirational research environment that enabled the young research team around Hasso Plattner and Alexander Zeier to generate valuable and new scientific insights into the complex world of enterprise computations. Even more than that, they developed groundbreaking innovations that will open the door to a new age, the age in which managers can base their decisions on complex computational real-time analysis of business data, and thus will change the way how businesses are being operated.

Prof. Donald Kossmann (ETH Zurich, Switzerland)

This is the first book on in-memory database systems and how this technology can change the whole industry. The book describes how to build in-memory databases:

what is different, what stays the same. Furthermore, the book describes how in-memory databases can become the single source of truth for a business.

Prof. Hector Garcia-Molina (Stanford University, California, USA)

Memory resident data can very significantly improve the performance of data intensive applications. This book presents an excellent overview of the issues and challenges related to in-memory data, and is highly recommended for anyone wishing to learn about this important area.

Prof. Hubert Oesterle (University of St. Gallen, Switzerland)

Technological innovations have again and again been enablers and drivers of innovative business solutions. As database management systems in the 1970s provided the grounds for ERP systems, which then enabled companies in almost all industries to redesign their business processes, upcoming in-memory databases will improve existing ERP-based business solutions (esp. in analytic processing) and will even lead to business processes and services being redesigned again. Plattner and Zeier describe the technical concepts of column- and row-based databases and encourage the reader to make use of the new technology in order to accomplish business innovation.

Prof. Michael Franklin (University of California at Berkeley, USA)

Hardware technology has evolved rapidly over the past decades, but database system architectures have not kept pace. At the same time, competition is forcing organizations to become more and more data-driven. These developments have driven a re-evaluation of fundamental data management techniques and tradeoffs, leading to innovations that can exploit large memories, parallelism, and a deeper understanding of data management requirements. This book explains the powerful and important changes that are brought about by in-memory data processing. Furthermore, the unique combination of business and technological insights that the authors bring to bear provide lessons that extend beyond any particular technology, serving as a guidebook for innovation in this and future Information Technology revolutions.

Prof. Sam Madden (Massachusetts Institute of Technology, Cambridge, USA)

Plattner and Zeier's book is a thorough accounting of the need for, and the design of, main memory database systems. By analyzing technology trends, they make a compelling case for the coming dominance of main-memory in database systems. They go on to identify a series of key design elements that main memory database system should have, including a column-oriented design, support for multi-core processor parallelism, and data compression. They also highlight several important requirements imposed by modern business processes, including heavy use of stored procedures and accounting requirements that drive a need for no-overwrite storage.

This is the first book of its kind, and it provides a complete reference for students and database designers alike.

Prof. Terry Winograd (Stanford University, California, USA)

There are moments in the development of computer technology when the ongoing evolution of devices changes the tradeoffs to allow a tectonic shift—a radical change in the way we interact with computers. The personal computer, the Web, and the smart phone are all examples where long-term trends reached a tipping point allowing explosive change and growth. Plattner and Zeier present a vision of how this kind of radical shift is coming to enterprise data management. From Plattner's many years of executive experience and development of data management systems, he is able to see the new space of opportunities for users—the potential for a new kind of software to provide managers with a powerful new tool for gaining insight into the workings of an enterprise. Just as the Web and the modern search engine changed our idea of how, why, and when we "retrieve information," large in-memory databases will change our idea of how to organize and use operational data of every kind in every enterprise. In this visionary and valuable book, Plattner and Zeier lay out the path for the future of business.

Prof. Warren B. Powell (Princeton University, Princeton, New Jersey, USA)

In this remarkable book, Plattner and Zeier propose a paradigm shift in memory management for modern information systems. While this offers immediate benefits for the storage and retrieval of images, transaction histories and detailed snapshots of people, equipment and products, it is perhaps even more exciting to think of the opportunities that this technology will create for the future. Imagine the fluid graphical display of spatially distributed, dynamic information. Or the ability to move past the flat summaries of inventories of equipment and customer requests to capture the subtle context that communicates urgency and capability. Even more dramatic, we can envision the real-time optimization of business processes working interactively with domain experts, giving us the information-age equivalent of the robots that make our cars and computers in the physical world today.

Prof. Wolfgang Lehner (Technical University of Dresden, Germany)

This book shows in an extraordinary way how technology can drive new applications—a fascinating journey from the core characteristics of business applications to topics of leading-edge main-memory database technology.

Industry

Heiko Hubertz (CEO and Founder, Bigpoint GmbH, Hamburg, Germany)

It rarely happens that an emerging new technology can be seen as a "game changer" for a whole industry. But I think this is the case when it comes to in-memory database systems. Tasks which took hours or days can now be executed in seconds which is the technical foundation for establishing completely new business ideas. It dramatically speeds up the execution time of data analytics and thereby creates a much greater experience for the end user on our Bigpoint Gaming Platform. The book "In-Memory Data Management—Technology and Applications" from Hasso Plattner and Alexander Zeier describes not only the technical foundations but also the implications for new exciting applications.

Dr. Ralf Schneider (CIO, Allianz SE, Munich, Germany)

Being IT savvy and leveraging advances in Information Technology is the most important competitive advantage in today's business world. I see in-memory technology as described in the book "In-Memory Data Management—Technology and Applications" from Plattner and Zeier as one of the most important innovations in the field of IT. The value for the economy as a whole far outweighs the effort of adapting the applications to this new standard. It dramatically speeds up the execution time of business processes, it allows for significantly more detailed and real-time data analytics and thereby creates a much greater experience for the end user. It brings business computing to a level where it is absolutely snappy and fun to work with.

Bill McDermott (Co-CEO, SAP AG, Newtown Square, Pennsylvania, USA)

We are witnessing the dawn of a new era in enterprise business computing, defined by the near instantaneous availability of critical information that will drive faster decision making, new levels of business agility, and incredible personal productivity for business users. With the advent of in-memory technology, the promise of real-time computing is now reality, creating a new inflection point in the role IT plays in driving sustainable business value. In their review of in-memory technology, Hasso Plattner and Alexander Zeier articulate how in-memory technology can drive down costs, accelerate business, help companies reap additional value out of their existing IT investments, and open the door to new possibilities in how business applications can be consumed. This work is a "must read" for anyone who leverages IT innovation for competitive advantage.

Falk F. Strascheg (Founder and General Partner, EXTOREL GmbH, Munich, Germany)

Since the advent of the Internet we have been witnessing new technologies coming up quickly and frequently. It is, however, rare that these technologies become innovations in the sense that there are big enough market opportunities. Hasso Plattner has proven his ability to match business needs with technical solutions more than once, and this time he presents the perhaps most significant innovation he has ever been working on: Real-Time Business powered by In-Memory Computing. As the ability for innovation has always been one of the core factors for competitiveness this is a highly advisable piece of reading for all those who aim to be at the cutting edge.

Gerhard Oswald (COO, SAP AG, Walldorf, Germany)

In my role as COO of SAP it is extremely important to react quickly to events and to have instant access to the current state of the business. At SAP, we have already moved a couple of processes to the new in-memory technology described in the book by Hasso Plattner and Alexander Zeier. I'm very excited about the recently achieved improvements utilizing the concepts described in this book. For example, I monitor our customer support messaging system everyday using in-memory technology to make sure that we provide our customers with the timely responses they deserve. I like that this book provides an outlook of how companies can smoothly adopt the new database technology. This transition concept, called the bypass solution, gives our existing customer base the opportunity to benefit from this fascinating technology, even for older releases of SAP software.

Hermann-Josef Lamberti (COO, Deutsche Bank AG, Frankfurt, Germany)

Deutsche Bank has run a prototype with an early version of the in-memory technology described in the book by Hasso Plattner and Alexander Zeier. In particular, we were able to speed up the data analysis process to detect cross-selling opportunities in our customer database, from previously 45 min to 5 s. In-memory is a powerful new dimension of applied compute power.

Jim Watson (Managing General Partner, CMEA Capital, San Francisco, California, USA)

During the last 50 years, every IT era has brought us a major substantial advancement, ranging from mainframe computers to cloud infrastructures and smart phones. In certain decades the strategic importance of one technology versus the other is dramatically different and it may fundamentally change the way in which people do business. This is what a Venture Capitalist has to bear in mind when identifying new trends that are along for the long haul. In their book, Hasso and Alex not only describe a market-driven innovation from Germany, that has the potential to change the enterprise software market as a whole, but they also present a working prototype.

Martin Petry (CIO, Hilti AG, Schaan, Liechtenstein)

Hilti is a very early adopter of the in-memory technology described in the book by Hasso Plattner and Alexander Zeier. Together with SAP, we have worked on developing prototypical new applications using in-memory technology. By merging the transactional world with the analytical world these applications will allow us to gain real-time insight into our operations and allow us to use this insight in our interaction with customers. The benefit for Hilti applying SAP's in-memory technology is not only seen in a dramatic improvement of reporting execution speed—for example, we were able to speed up a reporting batch job from 3 hours to seconds—but even more in the opportunity to bring the way we work with information and ultimately how we service our customers on a new level.

Prof. Norbert Walter (former Chief Economist of Deutsche Bank AG, Frankfurt, Germany)

Imagine you feel hungry. But instead of just opening the fridge (imagine you don't have one) to get hold of, say, some butter and cheese, you would have to leave the house for the nearest dairy farm. Each time you feel hungry. This is what we do today with most company data: We keep them far away from where we process them. In their highly accessible book, Hasso Plattner and Alexander Zeier show how in-memory technology moves data where they belong, promising massive productivity gains for the modern firm. Decision makers, get up to speed!

Paul Polman (CEO, Unilever PLC, London, UK)

There are big opportunities right across our value chain to use real-time information more imaginatively. Deeper, real time insight into consumer and shopper behavior will allow us to work even more closely and effectively with our customers, meeting the needs of today's consumers. It will also transform the way in which we serve our customers and consumers and the speed with which we do it. I am therefore very excited about the potential that the in-memory database technology offers to my business.

Tom Greene (CIO, Colgate-Palmolive Company, New York City, USA)

In their book, Hasso Plattner and Alexander Zeier not only describe the technical foundations of the new data processing capabilities coming from in-memory, but they also provide examples for new applications that can now be built on top. For a company like Colgate-Palmolive, these new applications are of strategic importance, as they allow for new ways of analyzing our transactional data in real time, which can give us a competitive advantage.

Dr. Vishal Sikka (CTO, Executive Board Member, SAP AG, Palo Alto, California, USA)

Hasso Plattner is not only an amazing entrepreneur, he is an incredible teacher. His work and his teaching have inspired two generations of students, leaders, professionals, and entrepreneurs. Over the last 5 years, we have been on a fantastic journey with him, from his early ideas on rethinking our core financials applications, to conceiving and implementing a completely new data management foundation for all our SAP products. This book by Hasso and Alexander, captures these experiences and I encourage everyone in enterprise IT to read this book and take advantage of these learnings, just as I have endeavored to embody these in our products at SAP.

To Annabelle and my family
AZ

Foreword

Is anyone else in the world both as well-qualified as Hasso Plattner to make a strong business case for real-time data analytics and describe the technical details for a solution based on insights in database design for Enterprise Resource Planning that leverage recent hardware technology trends?

The P of SAP has been both the CEO of a major corporation and a Professor of Computer Science at a leading research institute, where he and his colleagues built a working prototype of a main memory database for ERP. Taking advantage of rapid increases in DRAM capacity and in the number of the processors per chip, SanssouciDB demonstrates that the traditional split of separate systems for Online Transaction Processing (OLTP) and for Online Analytical Processing (OLAP) is no longer necessary for enterprise systems.

Business leaders now can ask ad hoc questions of the production transaction database and get the answer back in seconds. With the traditional divided OLTP/OLAP systems, it can take a week to write the query and receive the answer. In addition to showing how software can use concepts from shared nothing databases to scale across blade servers and use concepts from shared everything databases to take advantage of the large memory and many processors inside a single blade, this book touches on the role of Cloud Computing to achieve a single system for transactions and analytics.

Equally as important as the technical achievement, the "Bill Gates of Germany" shows how businesses can integrate this newfound ability to improve the efficiency and profitability of business. Moreover, if this ability is embraced and widely used, perhaps business leaders can quickly and finely adjust enterprise resources to meet rapidly varying demands so that the next economic downturn will not be as devastating to the world's economy as the last one.

Stanford University, CA, USA Prof. John L. Hennessy
University of California at Berkeley, CA, USA Prof. David A. Patterson

Preface

Preface to the Second Edition

About one year ago we published the first edition of this book. Within this last year, in-memory technology had such a big impact on the enterprise computing and application market that it truly marked an inflection point. This progress on the one hand, but also the resulting new questions on the other hand, convinced us that it is time for an extended second edition of our book.

The new content in the second edition targets the development and deployment of data-intensive applications that are designed for leveraging the capabilities of in-memory database systems. Among other new content, Sect. 6.1.1 introduces an in-memory application programming model that includes the most important aspects and guidelines for developing in-memory applications. To ease the tasks of application developers and database administrators, we discuss the graphical creation of database views in Sect. 6.1.5. Finally, we also elaborate on new features on application level, e.g., in Sect. 6.2.4 through the combination of data analytics and text search and by presenting two industry case studies in Sect. 9.2.

Of course, we could not have written a second edition of this book in such a short time without the help of our students at our Enterprise Platform and Integration Concepts chair. Therefore, we want to thank them in addition to the acknowledgement in the following preface of the first edition for their hard work and efforts.

Potsdam, 1 March 2012

Hasso Plattner
Alexander Zeier

Preface to the First Edition

We wrote this book because we think that the use of in-memory technology marks an inflection point for enterprise applications. The capacity per dollar and the availability of main memory has increased markedly in the last few years. This has led to a rethinking of how mass data should be stored. Instead of using mechanical disk drives it is now possible to store the primary data copy of a database in silicon-based main memory resulting in an orders of magnitude improvement in performance and allowing completely new applications to be developed. This change in the way data are stored is having, and will continue to have a significant impact on enterprise applications and ultimately on the way businesses are run. Having real-time information available at the speed of thought provides decision makers in an organization with insights that have, until now, not existed.

This book serves the interests of specific reader groups. Generally, the book is intended for anyone who wishes to find out how this fundamental shift in the way data is managed is affecting, and will continue to affect enterprise applications. In particular, we hope that university students, IT professionals, and IT managers, as well as senior management, who wish to create new business processes by leveraging in-memory computing, will find this book inspiring.

The book is divided into three parts:

- Part I gives an overview of our vision of how in-memory technology will change enterprise applications. This part will be of interest to all readers.
- Part II provides a more in-depth description of how we intend to realize our vision, and addresses students and developers, who want a deeper technical understanding of in-memory data management.
- Part III describes the resulting implications on the development and capabilities of enterprise applications, and is suited for technical as well as business-oriented readers.

Writing a book like this always involves more people than just the authors. We would like to thank the members of our Enterprise Platform and Integration Concepts group at the Hasso Plattner Institute at the University of Potsdam in Germany. Anja Bog, Martin Grund, Jens Krüger, Stephan Müller, Jan Schaffner, and Christian Tinnefeld are part of the HANA research group and their work over the last 5 years in the field of in-memory applications is the foundation for our book. Vadym Borovskiy, Thomas Kowark, Ralph Kühne, Martin Lorenz, Jürgen Müller, Oleksandr Panchenko, Matthieu Schapranow, Christian Schwarz, Matthias Uflacker, and Johannes Wust also made significant contributions to the book and our assistant Andrea Lange helped with the necessary coordination. Additionally, writing this book would not have been possible without the help of many colleagues at SAP. Cafer Tosun in his role as the link between HPI and SAP not only coordinates our partnership with SAP, but also actively provided sections for our book. His team members Andrew McCormick-Smith and Christian Mathis added important text passages to the book. We are grateful for the work of Joos-Hendrik Boese, Bernhard Fischer, Enno Folkerts, Andreas Herschel, Sarah Kappes,

Christian Münkel, Frank Renkes, Frederik Transier, and other members of his extended team. We would like to thank Paul Hofmann for his input and for his valuable help in managing our research projects with American Universities. The results we achieved in our research efforts would also not have been possible without the outstanding help of many other colleagues at SAP. We would particularly like to thank Franz Färber and his team for their feedback and their outstanding contributions to our research results over the past years. Many ideas that we describe throughout the book were originally Franz's, and he is also responsible for their implementation within SAP. We especially want to emphasize his efforts.

Finally, we want to express our gratitude to SAP CTO Vishal Sikka for his sponsorship of our research and his personal involvement in our work. In addition, we are grateful to SAP COO Gerhard Oswald and SAP Co-CEOs Jim Hagemann Snabe and Bill McDermott for their ongoing support of our projects.

We encourage you to visit the official website of this book. The website contains updates about the book, reviews, blog entries about in-memory data management, and examination questions for students.

no-disk.com

Potsdam, 1 February 2011   Hasso Plattner
                                                           Alexander Zeier

The Essence of In-Memory Data Management

Imagine you live in a major US city. Now, imagine that every time you want a glass of water, instead of getting it from the kitchen, you need to drive to the airport, get on a plane and fly to Germany, and pick up your water there. From the perspective of a modern CPU, accessing data which is in-memory is like getting water from the kitchen. Accessing a piece of data from the computer's hard disk is like flying to Germany for your glass of water. In the past the prohibitive cost of main memory has made the flight to Germany necessary. The last few years, however, have seen a dramatic reduction in the cost per megabyte of main memory, finally making the glass of water in the kitchen a cost effective and much more convenient option.

This orders-of-magnitude difference in access times has profound implications for all enterprise applications. Things that in the past were not even considered because they took so long, now become possible, allowing businesses concrete insight into the workings of their company that previously were the subject of speculation and guess work.

The in-memory revolution that we describe in this book is not simply about putting data into memory and thus being able to work with it "faster". We also show how the convergence of two other major trends in the IT industry: (a) the advent of multi-core CPUs and the necessity of exploiting this parallelism in software and (b) the stalling access latency for DRAM, requiring software to cleverly balance between CPU and memory activity; have to be harnessed to truly exploit the potential performance benefits. Another key aspect of the vision of in-memory data management that we present, is a change in the *way* data are stored in the underlying database. As we will see in the next section, this is of particular relevance for the enterprise applications that are our focus. The power of in-memory data management is in connecting all these dots.

In-Memory Data Management in Combination with Columnar Storage

Our experience has shown us that many enterprise applications work with databases in a similar way. They process large numbers of rows during their execution, but crucially, only a small number of columns in a table might be of interest in a particular query. The columnar storage model that we describe in this book allows only the required columns to be read while the rest of the table can be ignored. This is in contrast to the more traditional row-oriented model, where all columns of a table—even those that are not necessary for the result—must be accessed.

The columnar storage model also means that the elements of a given column are stored together. This makes the common enterprise operation of aggregation much faster than in a row-oriented model where the data from a given column are stored amongst the other data in the row.

Parallelization Across Multiple Cores and Machines

Single CPU cores are no longer getting any faster but the number of CPU cores is still expected to double every 18 months. This makes exploiting the parallel processing capabilities of multi-core CPUs of central importance to all future software development. As we saw above, in-memory columnar storage places all the data from a given column together in memory making it easy to assign one or more cores to process a single column. This is called vertical fragmentation.

Tables can also be split into sets of rows and distributed to different processors, in a process called horizontal fragmentation. This is particularly important as data volumes continue to grow and have been used with some success to achieve parallelism in data warehousing applications. Both these methods can be applied, not only across multiple cores in a single machine, but across multiple machines in a cluster or in a data center.

Using Compression for Performance and to Save Space

Data compression techniques exploit redundancy within data and knowledge about the data domain. Compression applies particularly well to columnar storage in an enterprise data management scenario, since all data within a column (a) have the same data type and (b) in many cases there are few distinct values, for example in a country column or a status column. In column stores, compression is used for two reasons: to save space and to increase performance.

Efficient use of space is of particular importance to in-memory data management because, even though the cost of main memory has dropped considerably, it is still relatively expensive compared to disk. Due to the compression within the columns, the density of information in relation to the space consumed is increased. As a result more relevant information can be loaded for processing at a time thereby increasing performance. Fewer load actions are necessary in comparison to row storage, where even columns of no relevance to the query are loaded without being used.

Conclusion

In-memory data management is not only a technology but a different way of thinking about software development: we must take fundamental hardware factors into account, such as access times to main memory versus disk and the potential parallelism that can be achieved with multi-core CPUs. Taking this new world of hardware into account, we must write software that explicitly makes the best possible use of it. On the positive side for developers of enterprise applications, this book lays the technological foundations for a database layer tailored specifically to all these issues. On the negative side, however, the database will not take care of all the issues on its own. Developers must understand the underlying layers of software and hardware to best take advantage of the potential for performance gains. The goal of this book is to help build such understanding.

Contents

Acronyms

ACID	Atomicity, Consistency, Isolation, Durability
ALU	Arithmetic Logic Unit
AMD	Advanced Micro Devices
API	Application Programming Interface
ASP	Application Service Provider
ATP	Available-to-Promise
BI	Business Intelligence
BWA	Business Warehouse Accelerator
CAPEX	Capital Expenditures
CBTR	Combined Benchmark for Transactions and Reporting
CC	Concurrency Control
CID	CommitID
ccNUMA	Cache-Coherent NUMA
CPU	Central Processing Unit
CRM	Customer Relationship Management
DBMS	Database Management System
DRAM	Dynamic Random Access Memory
ELT	Extract, Load and Transform
ERP	Enterprise Resource Planning
ETL	Extract, Transform and Load
FIFO	First In, First Out
FSB	Front Side Bus
I/O	Input/Output
IaaS	Infrastructure-as-a-Service
IMDB	In-Memory Database
IMS	Information Management System
IT	Information Technology
LIFO	Last In, First Out
LRU	Least Recently Used
MDX	Multidimensional Expressions
MIS	Management Information System

MOLAP	Multidimensional OLAP
MVCC	Multiversion Concurrency Control
NUMA	Non-uniform Memory Access
ODS	Operational Data Stores
OLAP	Online Analytical Processing
OLTP	Online Transaction Processing
OPEX	Operational Expenditures
OS	Operating System
PaaS	Platform-as-a-Service
PADD	Parallel Add
PCM	Phase Change Memory
PFOR	Patched Frame-of-Reference
QPI	Quick Path Interconnect
RAM	Random Access Memory
RDBMS	Relational Database Management System
RLE	Run-Length Encoding
ROLAP	Relational OLAP
SaaS	Software-as-a-Service
SCM	Supply Chain Management
SQL	Structured English Query Language
SID	Surrogate Identifier
SIMD	Single Instruction Multiple Data
SPEC	Standard Performance Evaluation Corporation
SQL	Structured Query Language
SRM	Supplier Relationship Management
SSB	Star Schema Benchmark
SSD	Solid State Drive
SSE	Streaming SIMD Extensions
TCO	Total Cost of Ownership
TLB	Translation Lookaside Buffer
TPC	Transaction Processing Performance Council
UMA	Uniform Memory Access
VM	Virtual Machine
XML	Extensible Markup Language

Introduction

Over the last 50 years, advances in Information Technology (IT) have had a significant impact on the success of companies across all industries. The foundation for this success is the successful leveraging of the strengths of IT systems in the rapid and accurate processing of repetitive tasks and the integration of these systems into the business processes of a company. This combination allows a more accurate and complete picture of an organization to be created. Another key aspect to this is that the speed at which IT systems have been able to create this picture means that it can be based on the most recent data available. This aspect has often been described and associated with the term "real-time" as it suggests that every change that happens within a company is instantly visible.

Significant milestones have been reached in the pursuit of this goal throughout the history of enterprise computing. Examples are the development of relational databases and the introduction of SAP's R/3 ERP (Enterprise Resource Planning) system. However, as the sheer volume of data that needs to be processed has increased the milestones have shifted. Currently, most of the data within a company are still distributed throughout a wide range of applications and stored in several disjoint silos. Creating a unified view of these data is a cumbersome and time-consuming process. Additionally, business analysis reports typically do not run directly on operational data, but on aggregated data from a data warehouse. Operational data is transferred into this data warehouse in batch jobs, which makes flexible, ad hoc reporting on the most up-to-date data impossible. As a consequence, company leaders have to make decisions based on data which are either out of data or incomplete. This is obviously not a true "real-time" solution.

We predict that this is about to change as hardware architectures continue the rapid evolution we have seen since the introduction of the microprocessor. This has become especially significant in the last decade where multi-core processors and the availability of large amounts of main memory at low cost are now enabling new breakthroughs in the software industry. It has become possible to store data sets of whole companies entirely in main memory, offering performance that is orders of magnitudes faster than traditional disk-based systems. Hard disks, the only remaining mechanical device in a world of silicon, will soon only be

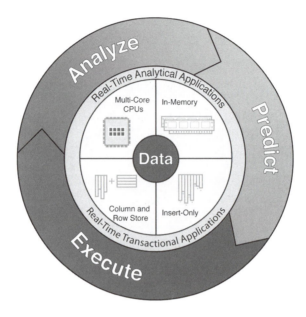

Fig. 1 Enterprise Performance In-memory Circle (EPIC)

necessary for backing up data. With in-memory computing and insert-only databases using row- and column-oriented storage, transactional and analytical processing can be unified. This development has the potential to change how enterprises work and finally offer the promise of real time computing.

As summarized in Fig. 1, the combination of the technologies mentioned above finally enables an iterative link between the instant analysis of data, the prediction of business trends, and the execution of business decisions without delays.

How can companies take advantage of in-memory applications to improve the efficiency and profitability of their business? We predict that this break through innovation will lead to fundamentally improved business processes, better decision-making, and new performance standards for enterprise applications across industries and organizational hierarchies. We are convinced that in-memory technology is a catalyst for innovation, and the enabler for a level of information quality that has not been possible until now. In-memory enterprise data management provides the necessary equipment to excel in a future where businesses face ever-growing demands from customers, partners, and shareholders. With billions of users and a hundred times as many sensors and devices on the Internet, the amount of data we are confronted with is growing exponentially. Being able to quickly extract business-relevant information not only provides unique opportunities for businesses; it will be a critical dfferentiator in future competitive markets.

With in-memory technology, companies can turn the massive amounts of data available into information to create strategic advantage. Operational business data can be interactively analyzed and queried directly by decision makers, opening up completely new scenarios and opportunities.

Consider the area of financial accounting, where data need to be frequently aggregated for reporting on a daily, weekly, monthly, or annual basis. With in-memory data management, the necessary filtering and aggregation can happen in real time. Accounting can be done anytime and in an ad hoc manner. Financial applications will not only be significantly faster, they will also be less complex and easier to use. Every user of the system will be able to directly analyze massive amounts of data. New data are available for analysis as soon as they are entered into the operational system. Simulations, forecasts, and what-if scenarios can be done on demand, anytime and anywhere. What took days or weeks in traditional disk-based systems can now happen in the blink of an eye. Users of in-memory enterprise systems will be more productive and responsive.

The concepts presented in this book describe new opportunities and areas for improvement across all industries. Below, we present a few examples:

- *Daily Operations*:
  Gain real-time insight into daily revenue, margin, and labor expenses.
- *Competitive Pricing*:
  Intuitively explore the impact of competition on product pricing to instantly understand the impact on profit contribution.
- *Risk Management*:
  Immediately identify high-risk areas across multiple products and services and run what-if scenarios on the fly.
- *Brand and Category Performance*:
  Evaluate the distribution and revenue performance of brands and product categories by customer, region, and channel at any time.
- *Product Lifecycle and Cost Management*:
  Get immediate insight into yield performance versus customer demand.
- *Inventory Management*:
  Optimize inventory and reduce out-of-stock scenarios based on live business events.
- *Financial Asset Management*:
  Gain a more up-to-date picture of financial markets to manage exposure to currencies, equities, derivatives, and other instruments.
- *Real-Time Warranty and Defect Analysis*:
  Get live insight into defective products to identify deviation in production processes or handling.

In summary, we foresee in-memory technology triggering improvements in the following three interrelated strategic areas:

- *Reduced Total Cost of Ownership*:
  The in-memory data management concepts described in this book enable the required analytical capabilities to be directly incorporated into the operational enterprise systems. Dedicated analytical systems can become a thing of the past.

This will allow enterprise systems to become less complex and easier to maintain, resulting in less hardware maintenance and fewer IT resource requirements.

- *Innovative Applications*:
  In-memory data management combines highvolume transaction processing capabilities with analytics, directly in the operational system. Planning, forecasting, pricing optimization, and other processes that previously had to be done on separate analytical systems can be dramatically improved and supported with new applications that were not possible before.

- *Better and Faster Decisions*:
  In-memory enterprise systems allow quick and easy access to information that decision makers need, providing them with new ways to look at the business. Simulation, what-if analyses, and planning can be performed interactively on operational data. Relevant information is instantly accessible. Collaboration within and across organizational units in a company is simplified. This can lead to a much more dynamic management style where problems can be dealt with as they happen.

At the research group, "Enterprise Platform and Integration Concepts", under the supervision of Prof. Dr. Hasso Plattner and Dr. Alexander Zeier at the Hasso Plattner Institute (HPI), we have been working since 2006 on research projects aimed at revolutionizing enterprise systems and applications. Our vision is that in-memory computing will enable completely new ways of operating a business and fulfill the promise of real-time data processing. This book serves to explain in-memory data management and how it is an enabler for this vision.

Part I
An Inflection Point
for Enterprise Applications

Chapter 1
Desirability, Feasibility, Viability:
The Impact of In-Memory

Abstract Sub-second respond time and real-time analytics are key requirements for applications that allow natural human computer interactions. We envision users of enterprise applications to interact with their software tools in such a natural way, just like any Internet user interacts with a web search engine today by refining search results on the fly when the initial results are not satisfying. In this initial chapter, we illustrate this vision of providing business data in real time and discuss it in terms of desirability, feasibility, and viability. We first explain the desire of supplying information in real time and review sub-second response time in the context of enterprise applications. We then discuss the feasibility based on in-memory databases that leverage modern computer hardware and conclude by demonstrating the economic viability of in-memory data management.

In-memory technology is set to revolutionize enterprise applications both in terms of functionality and cost due to a vastly improved performance. This will enable enterprise developers to create completely new applications and allow enterprise users and administrators to think in new ways about how they wish to view and store their data. The performance improvements also mean that costly workarounds, necessary in the past to ensure data could be processed in a timely manner, will no longer be necessary. Chief amongst these is the need for separate operational and analytical systems. In-memory technology will allow analytics to be run on operational data, simplifying both the software and the hardware landscape, leading ultimately to lower overall cost.

1.1 Information in Real Time: Anything, Anytime, Anywhere

Today's web search engines show us the potential of being able to analyze massive amounts of data in real time. Users enter their queries and instantly receive answers. The goal of enterprise applications in this regard is the same, but is barely reached. For example, call center agents or managers, are looking for specific pieces of information within all data sources of the company to better decide on products to offer

H. Plattner and A. Zeier, *In-Memory Data Management*,
DOI: 10.1007/978-3-642-29575-1_1, © Springer-Verlag Berlin Heidelberg 2012

to customers or to plan future strategies. Compared to web search with its instant query results, enterprise applications are slower, exposing users to noticeably long response times. The behavior of business users would certainly change if information was as instantly available in the business context as in the case of web search.

One major difference between web search and enterprise applications is the completeness of the expected results. In a web search only the hits that are rated most relevant are of interest, whereas all data relevant for a report must be scanned and reflected in its result. A web search sifts through an indexed set of data evaluating relevance and extracting results. In contrast, enterprise applications have to do additional data processing, such as complex aggregations . In a number of application scenarios, such as analytics or planning, data must be prepared before it is ready to be presented to the user, especially if the data comes from different source systems.

Current operational and analytical systems are separated to provide the ability to analyze enterprise data and to reach adequate query response times. The data preparation for analytics is applied to only a subset of the entire enterprise data set. This limits the data granularity of possible reports. Depending on the steps of preparation, for example, data cleansing, formatting, or calculations, the time window between data being entered into the operational system until being available for reporting might stretch over several hours or even days (see Sect. 7.1) for a more detailed discussion of the reasons, advantages, and drawbacks of the separation). This delay has a particular effect on performance when applications need to do both operational processing and analytics. Available-to-Promise (ATP), demand planning, and dunning applications introduced in Chap. 2 are examples of these types of applications. They show characteristics associated with operational processing as they must operate on up-to-date data and perform read and write operations. They also reveal characteristics that are associated with analytics like processing large amounts of data because recent and historical data is required. These applications could all benefit from the ability to run interactive what-if scenarios. At present, sub-second response times in combination with the flexible access to any information in the system are not available.

Figure 1.1 is an interpretation of information at the fingertips; a term coined by Bill Gates in 1994, when he envisioned a future in which arbitrary information is available from anywhere [58]. Our interpretation shows meeting attendees situated in several locations, all browsing, querying, and manipulating the same information in real time. The process of exchanging information can be shortened while being enriched with the potential to include and directly answer ad-hoc queries .

We now expand on the topics of sub-second response time, real-time analytics, and computation on the fly. These topics are vital for the scenario sketched above.

1.1.1 Response Time at the Speed of Thought

In web search, users query data using key word search. The meaning of key words may be ambiguous, which results in users redefining their search terms according to the received results. Sub-second response time is the enabler of such trial-and-

Fig. 1.1 Management meeting of the future [58]

error behavior. With sub-second response time in enterprise analytics, users would be able to employ the same method to query business data. Reporting queries could be defined and redefined interactively without long waiting periods to prepare data or create new reports.

The average simple reaction time for an observer to form a simple response to the presence of a stimulus is 220 ms [101]. Part of this time is needed to detect the stimulus and the remainder to organize the response. Recognition reaction time is longer because the process of understanding and comprehension has to take place. The average recognition reaction time has been measured to be 384 ms. Depending on the complexity of the context, the recognition reaction time increases. Assuming more complex contexts, the interval of 550–750 ms is what we call speed of thought. For trained users, who repeatedly perform the same action, the reaction times can be shorter than the above numbers and slow system response times will seem even longer for them.

Any interval sufficiently longer than the speed-of-thought interval will be detected as waiting time and the user's mind starts wandering to other topics, which is a process that cannot be consciously controlled. The longer the interval, the further the mind is taken off the task at hand. Sub-second response time is one step into the direction of helping the user to focus on a topic and to not become distracted by other tasks during waiting periods. Context switching between tasks is extremely tiring. Even for small tasks, the user has to find his way back to the topic and needs to remember what the next step was. If such a context switch can be omitted, the user's full attention is dedicated to exploring and analyzing data. The freedom to build queries on the results of previous ones without becoming distracted helps the user to dig deeper into data in a much shorter time.

Sub-second response time means that we can use enterprise applications in completely new ways, for example, on mobile devices. The expectation of mobile device users is that they will not have to wait more than a few seconds for a response from their device. If we are able to get a sub-second response time from our enterprise application then the response time for the mobile user, including the transfer time, will be within acceptable limits. This would allow managers to get the results of a dunning run on their phone while waiting for a flight. They could then call the worst debtors directly, hopefully leading to a rapid resolution of the problem. Compare this with traditional systems where a dunning run can potentially take hours.

1.1.2 Real-Time Analytics and Computation on the Fly

The basis for real-time analytics is to have all resources at disposal in the moment they are called for [142]. So far, special materialized data structures, called cubes, have been created to efficiently serve analytical reports. Such cubes are based on a fixed number of dimensions along which analytical reports can define their result sets. Consequently, only a particular set of reports can be served by one cube. If other dimensions are needed, a new cube has to be created or existing ones have to be extended. In the worst case, a linear increase in the number of dimensions of a cube can result in an exponential growth of its storage requirements. Extending a cube can result in a deteriorating performance of those reports already using it. The decision to extend a cube or build a new one has to be considered carefully. In any case, a wide variety of cubes may be built during the lifetime of a system to serve reporting, thus increasing storage requirements and also maintenance efforts.

   Instead of working with a predefined set of reports, business users should be able to formulate ad-hoc reports. Their playground should be the entire set of data the company owns, possibly including further data from external sources. Assuming a fast in-memory database, no more pre-computed materialized data structures are needed. As soon as changes to data are committed to the database, they will be visible for reporting. The preparation and conversion steps of data if still needed for reports are done during query execution and computations take place on the fly. Computation on the fly during reporting on the basis of cubes that do not store data, but only provide the interface for reporting, solves a problem that has existed up to now and allows for performance optimization of all analytical reports likewise.

1.2 The Impact of Recent Hardware Trends

Modern hardware is subject to continuous change and innovation, of which the most recent are the emergence of multi-core architectures and larger, less expensive main memory.[1] Existing software systems, such as database management systems, must be

---

[1] Main memory refers to silicon-based storage directly accessible from the CPU while in-memory refers to the concept of storing the primary data copy of a database in main memory. Main memory is volatile as data is lost upon power failure.

adapted to keep pace with these developments and exploit, to the maximum degree, the potential of the underlying hardware. In this section, we introduce database management systems and outline recent trends in hardware. We point out why those are key enablers for in-memory data management which makes true real-time analytics in the context of enterprise applications feasible.

1.2.1 Database Management Systems for Enterprise Applications

We define a Database Management System (DBMS) as a collection of programs that enable users to create and maintain a database [48]. The DBMS is a software system that facilitates the process of defining, constructing, manipulating, and sharing databases among various users and applications. It underpins all operations in an enterprise application and the performance of the enterprise application as a whole is heavily dependent on the performance of the DBMS.

Improving the performance of the database layer is a key aspect of our goal to remove the need for separate analytical systems and thus allow real-time access to all the data in an enterprise system. Here we describe the database concepts that are most relevant to enterprise applications and how hardware limitations at the database layer drove the decision to split Business Intelligence (BI) applications from the transactional system. We also discuss how recent developments in hardware are making it possible to keep all of an enterprise application's data in main memory resulting in significant performance gains.

The requirements of enterprise applications have been a key driver in the development of DBMSs, both in the academic and commercial worlds. Indeed, one of the earliest commercial DBMSs was the Information Management System (IMS) [119], developed in 1968 by IBM for inventory management in the National Aeronautics and Space Administration's Apollo space program. IMS was typical of the first commercial database systems in that it was built for a particular application.[2] At that time the effort required to develop complex software systems [19] meant that a DBMS that could provide a variety of general purpose functions to an application layer had not yet been developed. Creating new DBMSs to address each new enterprise application need was obviously time-consuming and inefficient; the motivation to have DBMSs that could support a wide range of enterprise applications was strong. Their development became a common goal of both research and industry.

1.2.1.1 Enterprise Applications and the Relational Model

The hierarchical data model used in early systems like IMS works well for simple transaction processing, but it has several shortcomings when it comes to analyzing the

---

[2] It should be noted that reengineered versions of IMS have subsequently come to be used in a wide variety of scenarios.

data stored in the database, a fundamental requirement for an enterprise application. In particular, combining information from different entity types can be inefficient if they are combined based on values stored in leaf nodes. In large hierarchies with many levels this takes a significant amount of time.

In 1970, Codd introduced a relational data model—and an associated relational calculus based on a set of algebraic operators—that was flexible enough to represent almost any dependencies and relations among entity types [27]. Codd's model gained quick acceptance in academic circles, but it was not until 1974—when a language initially called Structured English Query Language (SEQUEL) [21] was introduced, which allowed the formulation of queries against relational data in a comparatively user friendly language—that acceptance of the relational model became more widespread. Later, the description of the query language was shortened to Structured Query Language (SQL), and in conjunction with the relational model it could be used to serve a wide variety of applications. Systems based on this model became known as Relational Database Management Systems (RDBMS) (Sect. 5.1).

The next significant advance in RDBMS development came with the introduction of the concept of a transaction [61]. A transaction is a fixed sequence of actions with a well-defined beginning and a well-defined ending. This concept [73] coined the term ACID, which describes the properties of a transaction. ACID is an acronym for the terms Atomicity, Consistency, Isolation, and Durability. Atomicity is the capability to make a set of different operations on the database appear as a single operation to the user, and all of the different operations should be executed or none at all. The consistency property ensures that the database is in a consistent state before the start and after the end of a transaction. In order to ensure consistency among concurrent transactions, isolation is needed, as it fulfills the requirement that only a transaction itself can access its intermediate data unless the transaction is not finished. Atomicity, consistency, and isolation affect the way data is processed by a DBMS. Durability guarantees that a successful transaction persists and is not affected by any kind of system failure. These are all essential features in an enterprise application.

RDBMSs supporting ACID transactions provided an efficient and reliable way for storing and retrieving enterprise data. Throughout the 1980s, customers found that they could use enterprise applications based on such systems to process their operational data, so-called Online Transaction Processing (OLTP), and for any analytics or Online Analytical Processing (OLAP), they needed [170]. In the rest of the book we will use the terms OLTP and the processing of operational data interchangeably, and the terms OLAP and analytical processing interchangeably.

1.2.1.2 Separation of Transaction and Analytical Processing

As data volumes grew, RDBMSs were no longer able to efficiently service the requirements of all categories of enterprise applications. In particular, it became impossible for the DBMS itself to service ad-hoc queries on the entire transactional database in a timely manner.

Table 1.1 Access and read times for disk and main memory

Action	Time
Main memory access	100 ns
Read 1 MB sequentially from memory	250,000 ns
Disk seek	5,000,000 ns
Read 1 MB sequentially from disk	30,000,000 ns

One of the reasons the DBMS was unable to handle these ad-hoc queries is the design of the database schemas that underlie most transactional enterprise applications. OLTP schemas are highly normalized to minimize the data entry volume and to speed up inserts, update and deletes. This high degree of normalization is a disadvantage when it comes to retrieving data, as multiple tables may have to be joined to get all the desired information. Creating these joins and reading from multiple tables can have a severe impact on performance, as multiple reads to disk may be required. Analytical queries need to access large portions of the whole database, which results in long run times with regard to traditional solutions.

Online Analytical Processing (OLAP) systems were developed to address the requirement of large enterprises to analyze their data in a timely manner. These systems relied on specialized data structures [170] designed to optimize read performance and provide quick processing of complex analytical queries. Data must be transferred out of an enterprise's transactional system, into an analytical system and then prepared for predefined reports.

The transfer happens in cyclic batches, in a so-called Extract, Transform, and Load (ETL) process [181]. The required reports may contain data from a number of different source systems. This must be extracted and converted into a single format that is appropriate for transformation processing. Rules are then applied during the transformation phase to make sure that the data can be loaded into the target OLAP system. These rules perform a number of different functions, for example, removing duplicates, sorting and aggregation. Finally, the transformed data is loaded into a target schema optimized for fast report generation.

This process has the severe limitation in that one is unable to do real-time analytics as the analytical queries are posed against a copy of the data in the OLAP system that does not include the latest transactions.

1.2.1.3 Performance Advantage of In-Memory Technology over Disk

The main reason that current RDBMSs cannot perform the required queries fast enough is that query data must be retrieved from disk. Modern systems make extensive use of caching to store frequently accessed data in main memory but for queries that process large amounts of data, disk reads are still required. Simply accessing and reading the data from disk can take a significant amount of time. Table 1.1 shows the access and read times for disk and main memory (based on [40]).

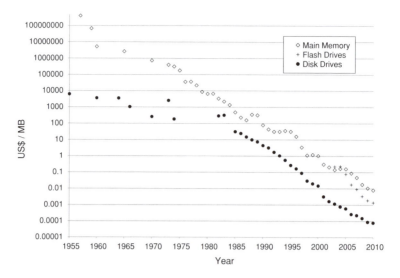

Fig. 1.2 Storage price development

Main memory or in-memory databases have existed since the 1980s [56] but it is only recently that Dynamic Random Access Memory (DRAM) has become inexpensive enough to make these systems a viable option for large enterprise systems. The ability of the database layer in an enterprise application to process large volumes of data quickly is fundamental to our aim of removing the need for a separate analytics systems. This will allow us to achieve our goal of providing a sub-second response time for any business query. In-memory databases based on the latest hardware can provide this functionality and they form the cornerstone of our proposed database architecture discussed in Chap. 3.

1.2.2 Main Memory is the New Disk

Since in-memory databases utilize the server's main memory as primary storage location, the size, cost, and access speed provided by main memory components are vitally important. With the help of data compression, today's standard server systems comprise sufficiently large main memory volumes to accommodate the entire operational data of all companies (Sect. 4.3). Main memory, as the primary storage location is, nevertheless, becoming increasingly attractive as a result of the decreasing cost/size ratio. The database can be directly optimized for main memory access, omitting the implementation of special algorithms to optimize disk access.

Figure 1.2 provides an overview of the development of main memory, disk, and flash storage prices over time. The cost/size relation for disks as well as main memory has decreased exponentially in the past. For example, the price for 1 MB of disk space dropped below US $ 0.01 in 2001, which is a rapid decrease compared to the cost of more than US $ 250 in 1970. A similar development can be observed for main memory. In addition to the attractiveness of fitting all operational business data of a company into main memory, optimizing and simplifying data access accordingly, the access speed of main memory compared to that of disks is four orders of magnitude faster: a main memory reference takes 100 ns [41]. Current disks typically provide read and write seek times about 5 ms [81, 35].

1.2.3 From Maximizing CPU Speed to Multi-Core Processors

In 1965, Intel co-founder Gordon E. Moore made his famous prediction about the increasing complexity of integrated circuits in the semiconductor industry [117]. The prediction became known as Moore's Law, and has become shorthand for rapid technological change. Moore's Law states that the number of transistors on a single chip is doubled approximately every two years [89].

In reality, the performance of Central Processing Units (CPUs) doubles every 20 months on average. The brilliant achievement that computer architects have managed is not only creating faster transistors, which results in increased clock speeds, but also in an increased number of transistors per CPU per square meter, which became cheaper due to efficient production methods and decreased material consumption. This leads to higher performance for roughly the same manufacturing cost. For example, in 1971, a processor consisted of 2,300 transistors whereas in 2006 it consisted of about 1.7 billion transistors at approximately the same price. Not only does an increased number of transistors play a role in performance gain, but also more efficient circuitry. A performance gain of up to a factor of two per core has been reached from one generation to the next, while the number of transistors remained constant.

Figure 1.3 provides an overview of the development of processor clock speed and the number of transistors from 1971 to 2010 based on [77, 88, 44]. As shown, the clock speed of processors had been growing exponentially for almost 30 years, but has stagnated since 2002. Power consumption, heat distribution and dissipation, and the speed of light have become the limiting factors for Moore's Law [121]. The Front Side Bus (FSB) speed, having grown exponentially in the past, has also stagnated.

In 2001, IBM introduced the first processor on one chip, which was able to compute multiple threads at the same time independently. The IBM Power 4 [13] was built for the high-end server market and was part of IBM's Regatta Servers. Regatta was the code name for a module containing multiple chips, resulting in eight cores per module [128]. In 2002, Intel introduced its proprietary hyper-threading technology, which optimizes processor utilization by providing thread-level parallelism on a single core. With hyper-threading technology, multiple logical processors with duplicated architectural state are created from a single physical processor. Several tasks

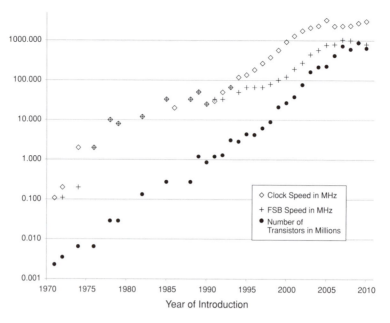

Fig. 1.3 Clock speed, FSB speed, and transistor development

can be executed virtually in parallel, thereby increasing processor utilization. Yet, the tasks are not truly executed in parallel because the execution resources are still shared and only multiple instructions of different tasks that are compatible regarding resource usage can be executed in a single processing step. Hyper-threading is applicable to single-core as well as to multi-core processors.

Until 2005, single-core processors dominated the home and business computer domain. For the consumer market, multi-core processors were introduced in 2005 starting with two cores on one chip, for example, Advanced Micro Devices's (AMD) Athlon 64 X2. An insight into the development of multi-core processors and future estimates of hardware vendors regarding the development of multi-core technology is provided in Fig. 1.4. At its developer forum in autumn 2006, Intel presented a prototype for an 80-core processor, while IBM introduced the Cell Broadband Engine with ten cores in the same year [70]. The IBM Cell Broadband Engine consists of two different types of processing elements, one two-core PowerPC processing element and up to eight synergistic processing elements that aim at providing parallelism at all abstraction levels. In 2008, Tilera introduced its Tile64, a multi-core processor for the high-end embedded systems market that consists of 64 cores [173]. 3Leaf is offering a product that is based on the HyperTransport architecture [3] with 192 cores. In the future, higher numbers of cores are anticipated on a single chip. In 2008, Tilera predicted a chip with 4,096 cores by 2017 for the embedded systems market and Sun estimated that servers are going to feature 32 and up to 128 cores by 2018 [18].

Fig. 1.4 Development of number of cores

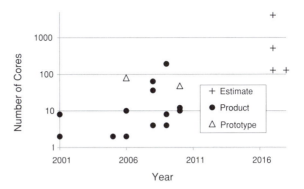

The developers of current and future software systems have to take multi-core and multi-processor machines into account. Software programs can no longer implicitly utilize the advances in processor technology, as was the case in the past decades with the growth of processor clock speed. In other words, "the free lunch is over" [168]. Utilization of parallelism through many processing units has to be explicitly incorporated into software by either splitting up algorithms across several computing units or executing many operations concurrently, even on the same set of data in case of a database system (Sect. 4.2). A programming paradigm, which scales well with an increasing number of single processing units, includes lock-free data structures and algorithms [167] (Sect. 5.6). With the development of multi-core architectures the stagnation of clock speed per core is compensated and technology advances as well as software development are heading in a new direction.

1.2.4 Increased Bandwidth Between CPU and Main Memory

The increased performance of the FSB, which so far has been the only interface from the CPU to main memory and all other input/output (I/O) components, that no longer keeps up with the exponential growth of processor performance anymore as can be seen in Fig. 1.3. Important for the calculation of the theoretical memory throughput are clock speed (cycles per second) and bandwidth of the data bus.

The increased clock speed and the use of multiple cores per machine are resulting in a widening gap between the ability of processors to digest data and the ability of the infrastructure to provide data. In-memory and column-oriented data storage enable the usage of additional processing power despite the bottleneck created by the aforementioned widening gap. High compression rates of column-oriented storage can lead to a better utilization of bandwidth. In-memory data storage can utilize enhanced algorithms for data access, for example, prefetching. We will discuss in-memory and column-oriented storage for database systems later in this book (Chap. 4).

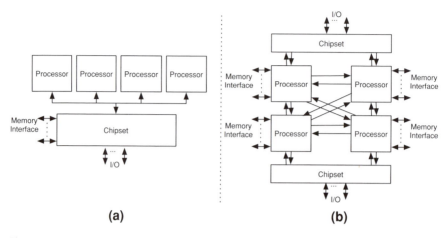

Fig. 1.5 **a** Shared FSB, **b** intel quick path interconnect [86]

Using compressed data and algorithms that work on compressed data is standard technology and has already proven to be sufficient to compensate the data supply bottleneck for machines with a small number of cores. It is, however, failing with the addition of many more cores. Experiments with column-oriented, compressed in-memory storage and data-intensive applications showed that the FSB was well utilized, though not yet congested, in an eight-core machine. The data processing requirements of the same applications on a 24-core machine surmounted the FSB's ability to provide enough data. From these experiments we can see that new memory access strategies are needed for machines with even more cores to circumvent the data supply bottleneck. Processing resources are often underutilized and the growing performance gap between memory latency and processor frequency intensifies the underutilization [37].

Figure 1.3 provides an overview of the development of the FSB speed. Intel improved the available transfer rate, doubling the amount of data that can be transferred in one cycle or added additional independent buses on multi-processor boards. The HyperTransport protocol was introduced by AMD in 2001 [3] to integrate the memory controller into the processor. Similar to the HyperTransport protocol, Intel introduced Quick Path Interconnect (QPI) [86] in the second half of 2008. QPI is a point-to-point system interconnect interface for memory and multiple processing cores, which replaces the FSB. Every processor has one or multiple memory controllers with several channels to access main memory in addition to a special bus to transfer data among processors. Compared to Intel FSB in 2007 with a bandwidth of 12.8 GB/s, QPI helped to increase the available bandwidth to 25.6 GB/s in 2008 [86]. In Intel's Nehalem EP chips, each processor has three channels from the memory controller to the physical memory [54]. In Intel's Nehalem EX chips, these channels have been expanded to four channels per processor [55].

Figure 1.5 gives an overview of the different architectures. In QPI, as shown in Fig. 1.5b, every processor has its exclusively assigned memory. On an Intel XEON

7560 (Nehalem EX) system with four processors, benchmark results have shown that a throughput of more than 72 GB/s is possible [55]. In contrast to using the FSB, shown in Fig. 1.5a, the memory access time differs between local memory (adjacent slots) and remote memory that is adjacent to the other processing units. As a result of this characteristic, architectures based on the FSB are called Uniform Memory Access (UMA) and the new architectures are called Non-Uniform Memory Access (NUMA). We differentiate between cache-coherent NUMA (ccNUMA) and non cache-coherent NUMA systems. In ccNUMA systems, all CPU caches share the same view to the available memory and coherency is ensured by a protocol implemented in hardware. Non cache-coherent NUMA systems require software layers to take care of conflicting memory accesses. Since most of the available standard hardware only provides ccNUMA, we will solely concentrate on this form.

To exploit NUMA completely, applications have to be made aware of primarily loading data from the locally attached memory slots of a processor. Memory-bound applications might see a degradation of up to 25 % of their performance if only remote memory is accessed instead of the local memory [55]. Reasons for this degradation can be the saturation of the QPI link between processor cores to transport data from the adjacent memory slot of another core, or the influence of higher latency of a single access to a remote memory slot. The full degradation might not be experienced, as memory caches and prefetching of data mitigates the effects of local versus remote memory. Assume a job can be split into many parallel tasks. For the parallel execution of these tasks distribution of data is relevant. Optimal performance can only be reached if the executed tasks solely access local memory. If data is badly distributed and many tasks need to access remote memory, the connections between the processors can become flooded with extensive data transfer.

Aside from the use for data-intensive applications, some vendors use NUMA to create alternatives for distributed systems. Through NUMA, multiple physical machines can be consolidated into one virtual machine. Note the difference in the commonly used term of virtual machine, where part of a physical machine is provided as a virtual machine. With NUMA, several physical machines fully contribute to the one virtual machine giving the user the impression of working with an extensively large server. With such a virtual machine, the main memory of all nodes and all CPUs can be accessed as local resources. Extensions to the operating system enable the system to efficiently scale out without any need for special remote communication that would have to be handled in the operating system or the applications. In most cases, the remote memory access is improved by the reservation of some local memory to cache portions of the remote memory. Further research will show if these solutions can outperform hand-made distributed solutions. 3Leaf, for example, is a vendor that uses specialized hardware. Other companies, for example, ScaleMP [120] rely on pure software solutions to build virtual systems. In summary, we have observed that the enhancement of the clock speed of CPU cores has tended to stagnate, while adding more cores per machine is now the reason for progress. As we have seen, increasing the number of cores does not entirely solve all existing problems, as other bottlenecks exist, for example, the gap between memory access speed and the clock speed of CPUs. Compression reduces the effects of this gap at the expense of computing cycles. NUMA as an alternative interconnection strategy for memory access through multiple cores has been developed. Increased performance through the addition of more cores and NUMA can only be utilized by adapting the software accordingly. In-memory databases in combination with column-oriented data storage are particularly well suited for multi-core architectures. Column-oriented data storage inherently provides vertical partitioning that supports operator parallelism (Sect. 4.2).

1.3 Reducing Cost Through In-Memory Data Management

In this section, we discuss the viability of using in-memory technology for enterprise data management and the financial costs of setting up and running an enterprise system. After an overview of the major cost factors, we look at how an architecture based on in-memory technology can help to reduce costs.

1.3.1 Total Cost of Ownership

The Total Cost of Ownership (TCO) is a business formula designed to estimate the lifetime costs of acquiring and operating resources, which in our case is an enterprise software system. The decision as to which hardware or software will be acquired and implemented will have a serious impact on the business. It is crucial to obtain an accurate cost estimate. The TCO measure was introduced when it became obvious that it is not sufficient to base IT decisions solely on the acquisition costs of the equipment because a substantial part of the cost is incurred later in the system lifecycle. The TCO analysis includes direct costs, for example, hardware acquisition, and indirect costs such as training activities for end users [96].

The primary purpose of introducing TCO is to identify all hidden cost factors and to supply an accurate and transparent cost model. This model can help to identify potential cost problems early. It is also a good starting point for a return on investment calculation or a cost-benefit analysis. These business formulas go beyond the TCO analysis and take the monetary profit or other benefits into consideration. Both are used to support decisions about technology changes or optimization activities.

Estimating TCO is a challenging exercise. Particular difficulties lie in creating an accurate cost model and estimating hidden costs. In many cases, the TCO analysis leads to a decision between a one-time investment on the one hand and higher ongoing costs on the other hand. This is known as the TCO tradeoff [39]. An investment in centralization and standardization helps to simplify operations and reduces overhead, thus reducing the cost of support, upgrades, and training. As another example, an initial investment in expensive high end hardware can boost system performance and facilitate development, administration, and all other operations.

1.3.2 Cost Factors in Enterprise Systems

The cost factors in setting up and maintaining an enterprise system include the cost of buying and operating the hardware infrastructure, as well as the cost of buying, administrating, and running the software.

1.3.2.1 Hardware Infrastructure and Power Consumption

The kind of hardware needed depends on the specific requirements of the applications. Enterprise applications of a certain complexity require high availability and low response time for many users on computationally complicated queries over huge amounts of data. The occurrence of many challenging requirements typically implies the need for high end hardware.

Power costs for servers and for cooling systems constitute a major part of the ongoing costs. The cost for power and the infrastructure needed for power distribution and cooling makes up about 30 % of the total monthly cost of running a large data center, while the server costs, which are amortized over a shorter time span of three years, constitute close to 60 %. This numbers refer to very large data centers (about 50,000 servers) and if their power usage efficiency is very high [75]. So, in many smaller data centers, the power and power infrastructure might contribute a greater portion of the cost.

1.3.2.2 System Administration

The cost impact of system administration is largely determined by the time and human effort it requires. These factors are determined by the performance of the system and the complexity of the tasks. The first tasks include the initial system setup and the configuration and customization of the system according to the business structure of the customer. Upgrades and extensions can affect every single component in the system, so it is a direct consequence that they get more expensive as the system gets more complex. The same is true for the monitoring of all system components in order to discover problems (like resource bottlenecks) early. The costs and the risks involved in scaling and changing a complex system are limiting factors for upgrading or extending an existing system.

Typical administrative tasks involve scanning, transforming, or copying large amounts of data. As an example, when a new enterprise system is set up, it is often necessary to migrate data from a legacy system. During system upgrades and extensions, complete tables might need to be transformed and reformatted. For system backups as well as for testing purposes complete system copies are created.

1.3.3 In-Memory Performance Boosts Cost Reduction

If all data can be stored in main memory instead of on disk, the performance of operations on data, especially on mass data, is improved. This has impact on every aspect of the system: it affects the choices of hardware components, but also the software architecture and the basic software development paradigms. Many performance crutches, like redundant materialized data views, have been introduced solely to optimize response time for analytical queries. The downside is that redundancy has to be maintained with significant overhead (Sect. 7.1). This overhead becomes superfluous when the system is fast enough to handle requests to scan huge amounts of data on the fly. Pushing data-intensive operations into the database simplifies the application software stack by fully utilizing the functionality of the database and by avoiding the necessity of transporting massive amounts of data out of the database for computation. If data migration and operations on mass data are accelerated, this automatically makes system copying, backups, archiving, and upgrade tasks less time consuming, thus reducing cost.

The improved performance of operations on mass data not only facilitates analytical queries, but also many of the day-to-day administrative tasks mentioned above. We have indicated that a simplification of a software system's architecture has multiple positive effects on the cost. In short, a simpler system, for example, an architecture with fewer layers and fewer components, is less expensive and faster to set up, easier to operate, easier to scale and change, and it generates fewer failures.

Our current experience in application development on the basis of in-memory technology shows that the size of the application code can be reduced up to 75 %. All the orchestration of layers, for example, caches and materialization, is not required any longer and algorithms are pushed down to the database to operate close to the data needed.

With respect to user-system interaction, improved performance and reduced complexity directly translate into reduced cost. The increased cost through wasted time that occurs whenever someone has to wait for a response from the system or has to invest time to understand a complex process affects all stages in the software lifecycle, from software development and system administration to the end user of the business application.

Compared to high-end disk-based systems that provide the necessary performance for database computing, the initial costs of in-memory systems, as well as running costs, are not that different. High-end disk-based systems provide the necessary bandwidth for computation through redundancy of disks. Although disks are able to provide more space less expensively, scaling along the number of disks and managing them is mandatory for comparable performance to in-memory computing.

In summary, we can say that using in-memory technology can help to reduce the TCO of an enterprise application by reducing the complexity of the application software layers, performing data-intensive tasks close to the data source, speeding up response times allowing users to make more efficient use of their time, as well as speeding up administrative tasks involving mass data.

1.4 Conclusion

In-memory and multi-core technology have the potential to improve the performance of enterprise applications and the value they can add to a business. In this chapter we described the potential impact of in-memory technology on enterprise applications and why it is the right time to make the switch.

We first identified desirable new features that can be added to enterprise applications if performance is improved. Chief among them was the ability to perform analytics on transactional data rather than having to use a separate BI system for analysis.

Having identified these features, we then looked at the feasibility of providing them in an enterprise application by using a redesigned DBMS based on in-memory technology. We showed how current trends in hardware can be utilized to realize our vision and how these new hardware developments complement our database design.

Finally, we discussed the viability of our approach by assessing the impact that using a DBMS based on in-memory technology has on the TCO of an enterprise system. We discussed how the better performance possible from such a system could reduce the TCO.

Chapter 2
Why Are Enterprise Applications So Diverse?

Abstract Today, even small businesses operate in different geographical locations and service different industries. This can create a number of challenges including those related to language, currencies, different regulatory requirements, and diverse industry expectations. For large organizations with a wider reach, the challenges are even greater. As organizations grow, they also need to keep track of huge amounts of information across different business areas. Modern enterprise applications need to be able to cope with these demands in a timely manner. In this chapter we provide a brief introduction to modern enterprise applications, describing selected tasks they perform and the business areas that they cover.

Enterprise applications are software systems that help organizations to run their businesses. They can add a degree of automation to the implementation of business processes as well as supporting tasks such as planning, data analysis, and data management. A key feature of an enterprise application is its ability to integrate and process data from different business areas, providing a holistic, real-time view of the entire enterprise.

Ideally, an enterprise application should be able to present all relevant information for a given context to the user in a timely manner, enabling effective decision making and allowing business departments to optimize their operations. This differentiates enterprise applications from other business software like spreadsheets, which are unable to pull data automatically from all the relevant data sources. Another factor that distinguishes enterprise applications from other types of software is that they are used exclusively in a business setting.

In this chapter, we describe the scope of an integrated enterprise application and resulting requirements (Sect. 2.1). Section 2.3 presents some selected enterprise application examples. The architecture of enterprise applications is discussed in Sect. 2.4. Resulting data access patterns are described in Sects. 2.5 and 2.6 closes this chapter.

H. Plattner and A. Zeier, *In-Memory Data Management*,
DOI: 10.1007/978-3-642-29575-1_2, © Springer-Verlag Berlin Heidelberg 2012

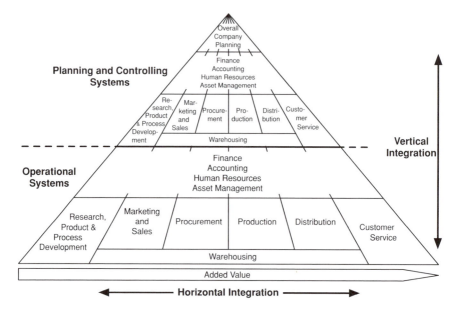

Fig. 2.1 Integrated enterprise information processing (based on [112] and [188])

2.1 Current Enterprise Applications

To get a sense of the variety of functionality that an integrated enterprise system has to cover, we present a view on an enterprise with regard to its divisions in Fig. 2.1. An enterprise system is in charge of integrating business processes that span multiple divisions (horizontal integration). It connects planning and controlling systems with operational systems (vertical integration).

Enterprise applications not only integrate information processing in homogeneous companies, but also cater to a wide range of different customer groups. For example, SAP Business Suite, the leading integrated Enterprise Resource Planning (ERP) package, is used within a large number of industries ranging from Aerospace and Defense, Telecommunications, and Banking to Industrial Machinery and Components Manufacturing. This leads to a situation of diverse and sometimes contradicting customer demands. For example, the product development and go-to-market lifecycle in the high-tech industry is less than six months, whereas in the chemical, railway, or oil and gas industry such a lifecycle can amount to decades [50, 161]. Companies in each of these diverse industries expects to enterprise system to be tailored for their requirements. To cope with this situation, successful enterprise applications have to allow parameterizations and the creation of industry-specific solutions. This includes giving customers the ability to determine how to query their data and enabling them to specify extensions to the underlying database tables if they need to include information that is not currently stored in the system. Another important customization feature in enterprise software is the ability to adapt existing business processes to customer-specific requirements.

International companies work across multiple time zones and rely on their software being available at all times. No downtime is acceptable. Enterprise applications must meet this requirement. This includes providing mechanisms for recovering from a wide range of failures, including power failures, data loss, and user input failures. Enterprise applications that are used in a number of different countries have to be internationalized to allow users to interact with them in their own language. They have to be compliant with all relevant national and international regulations and laws. This means that the software should be able to be easily updated because laws change on a regular basis, for example, with regard to income tax structures and tax rates. It is also often necessary for a company to maintain historical data for a number of years. The enterprise application should support this but should also make sure that the ever-growing data volume does not impact the processing of current data.

2.2 Examples of Enterprise Applications

We present below a subset of enterprise applications a company might use. We describe them in more detail throughout the rest of this section to provide the non-business reader with a sense of what enterprise applications do. For illustration, our company of reference has the following use cases:

- Forecast of product demand is needed for the following planning horizons. The forecast will be used as a basis for more detailed production planning (Demand Planning).
- Customers contact the company and order products (Sales Order Processing).
- Availability of the requested products is checked with every new order, and customers are informed if, and when, products can be shipped (Available-to-Promise).
- Some customers fall behind on payments for products received. Therefore, payments must be tracked and reminders generated for outstanding invoices (Dunning).
- The company performs a sales analysis each week (Sales Analysis).

Innovative enterprise applications that were not possible until now will be presented in Sect. 6.2.

Demand Planning

Demand planning is used to estimate future sales by combining several data sources. These include current and previous sales, new product introductions, product discontinuations, market forecasts, and other events with possible impact on buying behavior. The resulting outcome influences production planning, which determines the quantity and schedule for product manufacturing.

A demand planning application has the following characteristics: a single demand planning run for the company involves a large amount of data (up to 100 GB in larger companies). The data is created by hundreds of thousands of different products, their respective variations and configurations, and the fine-grained timely planning levels that allow planning on a daily or hourly basis.

The main operations on that data are aggregation and disaggregation because with every planning run, all current numbers are aggregated. Then, changes on a higher level in the planning hierarchy are applied, which are disaggregated to the finest granular level. The operation usually involves far more read operations than write operations, since all numbers are read, but only a fraction of them is actually changed. The underlying database system must be able to perform these operations with sub-second response time, as a human planner continuously works with the system in an interactive manner. Due to multi-user concurrency, isolated user contexts must be created to guarantee a consistent view on the data while planning. This acts as a multiplier for the overall storage and computing requirements.

We can see that this application has analytical aspects, but that it also requires real-time access to operational data. A combined transactional and analytical solution is required here.

Sales Order Processing

The main purpose of sales order processing is to capture customer orders. A sales order consists of a general section including customer name, billing address, and payment method. Details about the ordered products are stored as line items. Sales order processing involves read and write operations on transactional data. Read operations are required to check availability, item location and price. They are simple read operations, while write operations insert new sales orders and corresponding line items. More read operations than write operations occur and all transactions involve small amounts of data processed by highly predictable queries. Sales order processing has to have a fast response time, since clerks use the application interactively.

Available-to-Promise

The decision to accept an order for a product depends on the successful completion of an Available-to-Promise (ATP) check. The ATP application determines whether sufficient quantities of the product will be available in current and planned inventory levels for the requested delivery date [113]. To support flexible operations such as reordering where a high priority order overrides previously accepted low priority orders, ATP must aggregate across pending orders, rather than just materializing inventory levels. ATP encompasses read and write operations on large sets of data. Read operations dominate and are utilized to aggregate the respective time series. Write operations work on the fine-granular transactional level, such as to declare products as promised to customers. Variance in data types and records is small. The data volume grows only by adding new products, customers or locations.

Dunning

Dunning is the process of scanning through open invoices to identify those that are overdue, generating reminder notices for corresponding clients, and tracking which notices have been sent. It involves read operations on large amounts of transactional data. Either the read operations are batch-oriented, as the list of open items gets processed customer by customer, or the read operations are analytical queries, resulting in data aggregation, for example, when the overdue invoices for a selected customer group in a certain fiscal year must be determined. Read operations access very few attributes of the open items. Inserts are needed to keep track of issued payment reminders. Updates are needed to modify a customer's master data, such as the current dunning level, after a dunning run.

Sales Analysis

Sales analysis provides an overview of historical sales numbers. Usually, the analysis is used for decision making within an organization. The process involves read-only operations on large amounts of transactional and pre-aggregated data. For example, a sales manager might analyze sales volume by product group and country and compare it to previous years' numbers. Sales analysis comprises recent and historical data. The queries are multidimensional and usually scan and aggregate large amounts of data.

2.3 Enterprise Application Architecture

Current enterprise applications comprise client devices, application servers, and a persistency layer. Different enterprise application vendors implemented the business logic at different layers. For example, PeopleSoft preferred to concentrate the business logic at client devices whereas Oracle relies on stored procedures close to the database in the persistency layer to implement business functionality. SAP took a third option of having a thin client while all the business logic resides at scalable application servers. SAP's approach is shown schematically in Fig. 2.2.

In reality, the given figure would be more complicated, for example, as there is often redundancy added to the system at the database management system and application servers to increase performance. When this model was introduced with SAP R/3 about 20 years ago, the underlying rationale was that the database server was the limiting factor and as much load as possible should be shifted from the database server to multiple application servers. Usually multiple database management systems are supported by enterprise applications resulting in the situation that only

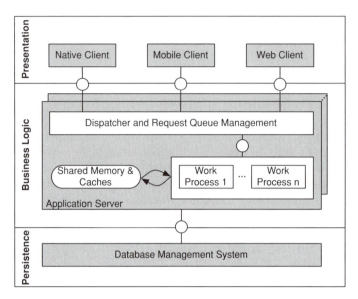

Fig. 2.2 Three-tier architecture of SAP R/3

the subset of all supported systems can be used in application development. This has led to a situation in which the application servers have become cumbersome and complex. One of the reasons is that application caches were introduced to store intermediate database results and all the computation is conducted in this tier to minimize database utilization.

2.4 Data Processing in Enterprise Applications

The data processed by an enterprise application comes in a number of different forms. Our customer analysis revealed that, for example, a medium-sized enterprise system today contains 100 GB of transactional data and 1 TB of read-only analytical data. Values for a large company are more than 35 TB of transactional data and 40 TB of read-only analytical data.

Processing this amount of data for planning and controlling purposes or to create reports required by management can have a significant impact on the performance of the overall system [124]. To alleviate these effects, most current enterprise applications separate these tasks into BI applications on the one hand and day-to-day transaction processing applications on the other hand. BI applications are characterized by read-intensive operations over large portions of the entire data in an enterprise system. This can comprise predefined reports run on a regular basis or complex ad-hoc queries—often requiring to aggregate large amounts of data. These queries can be in response to management questions formulated with regard to a current problem or an interesting aspect within a company, such as the current development of the company in a specific country or region of the world. These types of queries do not perform well on traditional transactional database schemas that are optimized for writing. This means that analyzing data stored in a number of different tables requires many tables to be joined, with a severe impact on performance.

BI applications tackle these problems by pre-aggregating data and storing the results in special read-optimized schemas. Analysis if this data is only possible along pre-selected dimensions [170]. In these read-optimized schemas certain attributes are duplicated across numerous tables to reduce the number of joins required. The pre-aggregated data is updated regularly by extracting data from the transactional system, transforming it according to the aggregation rules and the read-optimized schema, and then loading it into the BI application [93].

This separation of enterprise systems into those that perform OLTP and those that perform OLAP suggests a precise division of enterprise applications into transaction processing applications and analytical applications. This is far from the truth, since the previously discussed applications and others perform transaction processing and analytical tasks.

2.5 Data Access Patterns in Enterprise Applications

For the impact on the underlying in-memory data management system, it is relevant to distinguish between processing one single instance of an enterprise entity and the processing of attributes of a set of instances. A single instance is, for example, a sales order or a single customer. Set processing happens, for example, when all overdue invoices are read, or the top ten customers by sales volume are shown. Although transactional systems work on single instances of objects, most of the data in an integrated enterprise application is consumed by set processing [99].

Enterprise applications span operational systems as well as planning and controlling systems (Fig. 2.1). We can conclude that enterprise applications have a mixed workload [100].

2.6 Conclusion

In this chapter, it is evident that enterprise applications perform a very wide range of tasks and process large amounts of data. These data volumes are placing limits on the timeliness of data available for analysis, resulting in the need for separate analytical systems. This division of enterprise applications into two separate systems, which was conducted because of technical reasons, has a number of disadvantages. Therefore, some of the core themes that we focus on in this book are the technologies and techniques that can be employed to remove the need for separate analytical systems. These improvements allow enterprise application users real-time access to all the data in their enterprise system and to have a single source of truth.

Chapter 3
SanssouciDB: Blueprint for an In-Memory Enterprise Database System

Abstract This chapter provides an overview of how we plan to realize our vision of making real-time information about a business available in real time. We show how using an in-memory database, designed specifically for enterprise applications, provides significant performance advantages over general purpose disk-based DBMSs. Throughout the chapter we describe the technologies we use and why these are the best choices for our particular purposes.

The motto for the last 25 years of commercial DBMS development could well have been "One Size Fits All" [165]. Traditional DBMS architectures have been designed to support a wide variety of applications. These applications often have different characteristics and place different demands on the data management software making it impossible to optimize a DBMS for a specific task. In the main, the general-purpose database management systems that dominate the market today do everything well but do not excel in any area. In contrast major performance gains have been reported from database systems tailored to specific application areas, such as text search and text mining, stream processing, and data warehousing [165]. Directly incorporating the characteristics of these application areas and addressing them in the system architecture and in the data layout has been shown to improve performance significantly.[1]

As we emphasized in the previous chapters, our vision is to unite operational processing and analytical processing in one database management system for enterprise applications. We believe that this effort is an essential prerequisite to address the shortcomings of existing solutions to enterprise data management and to meet the requirements of tomorrow's enterprise applications [135].

We will illustrate our ideas with the help of SanssouciDB—our prototype database system for unified analytical and transaction processing.[2] SanssouciDB is a characteristic-oriented database system: it is specifically tailored to enterprise appli-

---

[1] In the following, systems that are tailored to the characteristics of a particular application area will be referred to as characteristic-oriented database systems.

[2] The concepts of SanssouciDB build on prototypes developed at the HPI and on an existing SAP database system.

H. Plattner and A. Zeier, *In-Memory Data Management*,
DOI: 10.1007/978-3-642-29575-1_3, © Springer-Verlag Berlin Heidelberg 2012

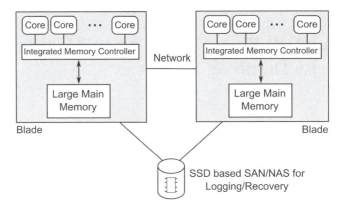

Fig. 3.1 SanssouciDB physical architecture

cations. We provide a number of examples of how the requirements of enterprise applications have influenced design decisions in the system, in particular we focus on the target hardware, the architecture, and the data store of SanssouciDB.

3.1 Targeting Multi-Core and Main Memory

Ideally, we would like to fit the complete database of an enterprise onto a single blade, that is, into a machine with a single main board containing multiple CPUs and a large array of main memory modules. This would simplify the design of SanssouciDB. However, not even the largest blades available at the time of writing allow us to do this. We need to utilize a cluster of multiple blades, where the blades are interconnected by a network (Fig. 3.1).

A necessary prerequisite for a database system running on such a cluster of blades, is a data partitioning strategy and the ability to distribute data across blades. Managing data across multiple blades introduces more complexity into the system. For example, distributed query processing algorithms that can access partitions in parallel across blades have to be implemented. Furthermore, accessing data via the network incurs higher communication costs than blade-local data access. Finally, different data partitioning strategies have an impact on query performance and load balancing. Therefore, from time to time, it can become necessary to reorganize the partitions to achieve better load balancing or to adapt to a particular query workload.

On the other hand, there are also substantial advantages to the cluster architecture: a single-blade approach results in a single point of failure, whereas the multi-blade approach enables high availability. If a blade unexpectedly fails, a stand-by blade can take over and load the data of the failed blade from external storage into its main memory. If no stand-by blade is available, the data can be loaded into the main memory of the remaining blades in the cluster (see Sect. 5.9 for a discussion on high availability). Furthermore, a multi-blade approach allows scale-out in cases where the existing number of blades cannot handle a given workload or data volume.

In cases such as these more blades can be added, the data can be further distributed, and a higher degree of data parallelism can be achieved (see Sect. 4.2 for a discussion on parallel data processing and scalability).

After choosing a multi-blade system as the target hardware, the next question is: should many less powerful, low-end blades be used or do we design our system to run on a small number of more powerful, high-end blades? For SanssouciDB, we chose the latter option, for two reasons: high-end blades are more reliable (see Sect. 5.9) and secondly, a smaller number of blades, each with the ability to manage a higher volume of data, allows more blade-local data processing. In our target hardware configuration, a typical blade has 2 TB of main memory and up to 64 cores. With 25 of these blades, we can hold the enterprise data of the largest companies in the world.

SanssouciDB adopts aspects of both the shared-nothing and shared-memory architectural approaches: an instance of SanssouciDB runs on each blade and each of these instances is responsible for a certain partition of the data. Therefore, from a global standpoint, we characterize the system as shared-nothing. Within a single blade, however, multiple cores have access to the data stored in the blade's memory. Hence, locally, SanssouciDB is a shared-memory system (see Sect. 4.2). Note that, as shown in Fig. 3.1, all blades are connected to the same non-volatile storage, for example, a Network Attached Storage (NAS) or a Storage Area Network (SAN). Therefore, one might even want to classify the architecture as shared disk. However, for SanssouciDB, the primary data copy is of importance and, since it is held in main memory and not on non-volatile storage, we classify the system as a whole as shared nothing.

To make efficient use of this architecture, SanssouciDB exploits parallelism at all levels. This includes distributed query processing (among blades), parallel query processing algorithms (among cores on a blade) and even taking advantage of special Single Instruction Multiple Data (SIMD) instructions at processor level. The subject of parallelism targeting this architecture is discussed in detail in Sect. 4.2 and also in Sect. 5.7, where we describe two important distributed and parallel query processing algorithms used in SanssouciDB.

An important factor to take into account when designing an in-memory system are the different latencies of the on-chip cache and main memory. Sect. 4.1 provides a detailed description of the effects of the memory hierarchy on in-memory data management. Finally, to help improve performance, we use flash-based storage solutions rather than conventional hard disks as our non-volatile storage.

3.2 Designing an In-Memory Database System

The relational model, introduced briefly in Chap. 1, is used by nearly all enterprise applications rely on the relational data model so we have made SanssouciDB a relational database system. The relations stored in SanssouciDB reside permanently in main memory; since access time to main memory is orders of magnitude faster than

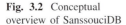

Fig. 3.2 Conceptual overview of SanssouciDB

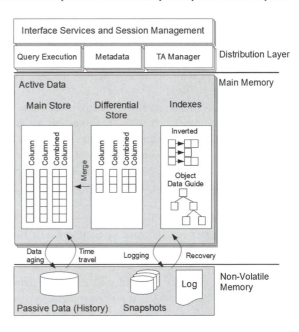

that for disk-resident database systems (Table 1.1). In-Memory Database Systems (IMDBs) [56] remove the main performance bottleneck for read operations found in a disk-based database system.

Figure 3.2 presents a conceptual overview of SanssouciDB. As described above, SanssouciDB is designed to run on a cluster of blades. A server process runs on each blade, the server process itself can run and manage multiple threads, one per physical core available on the blade. A scheduler (not shown in Fig. 3.2) is responsible for assigning work packages to the various threads (see Sect. 5.10).

To communicate with clients and other server processes, a SanssouciDB server process has an interface service and a session manager. The session manager keeps track of client connections and the associated parameters, for example, the connection timeout. The interface service provides the SQL interface and support for stored procedures (see Sect. 5.1). The interface service also implements the Application Programming Interface (API) to forward incoming requests to other client and server processes. The services run on top of the distribution layer, which is responsible for coordinating distributed metadata handling, distributed transaction processing, and distributed query processing. To allow fast, blade-local metadata lookups, the distribution layer replicates and synchronizes metadata across the server processes running on the different blades. The metadata contains information about the storage location of tables and their partitions. Because data is partitioned across blades, SanssouciDB provides distributed transactions and distributed query processing. The distribution layer also includes the transaction manager, labeled in the figure as TA Manager.

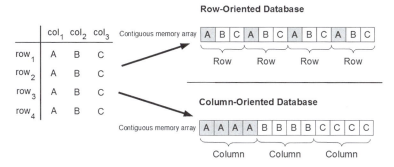

Fig. 3.3 Row- and column-oriented data layout

In SanssouciDB active data is kept in main memory. This primary copy of the database consists of a main store, a differential store, and a collection of indexes (see Sect. 3.3). Passive data is kept in non-volatile memory and is required to guarantee the persistence of the database.

While the issues concerning the distribution layer are interesting and far from trivial, the data interface services, session management, and distribution techniques in SanssouciDB have been developed in the context of classic disk-based. We omit a detailed discussion of these topics with the exception of distributed query processing, which is discussed in Sect. 4.2. In Part II of this book, we focus on the aspects of efficient in-memory data organization and access methods. The following section gives a brief introduction and provides pointers to the sections in the book that provide more details.

3.3 Organizing and Accessing Data in SanssouciDB

Traditionally, the data values in a database are stored in a row-oriented fashion, with complete tuples stored in adjacent blocks on disk or in main memory [38]. This can be contrasted with column-oriented storage, where columns, rather than rows, are stored in adjacent blocks. This is illustrated in Fig. 3.3.

Row-oriented storage allows for fast reading of single tuples but is not well suited to reading a set of results from a single column. The upper part of Fig. 3.4 exemplifies this by illustrating row and column operations in a row store.

Analysis of database accesses in enterprise applications has shown, however, that these set-based reads are the most common type of operation [135], making row-oriented databases a poor choice for enterprise applications. Column-oriented databases [2, 1], on the other hand, that store data in columns rather than in rows, are well suited to set-based operations. In particular, column scans, where all the data values that need to be scanned are located sequentially, can be implemented very efficiently. The lower part of Fig. 3.4 illustrates these considerations. Good scan performance make column-oriented databases a good choice for analytical process-ing; indeed, many commercial column-oriented DBMSs target the analytics market.

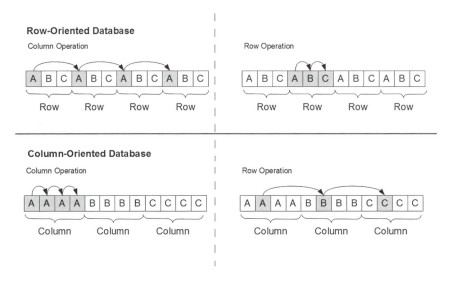

Fig. 3.4 Operations on row- and column-oriented databases

Examples are SAP Business Warehouse Accelerator and Sybase IQ. The disadvantage of these systems is that the performance of row-based operations, like the reading or writing of a complete tuple, is typically poor. To combine the best of both worlds, SanssouciDB allows certain columns to be stored together, such as columns that are frequently queried as a group (see Sect. 4.4). In the following, we refer to these groups of columns as combined columns. Allowing these column types combines the advantage of the column-oriented data organization to allow for fast reads with good write performance.

Even though main memory sizes have grown rapidly, data compression techniques are still necessary if we are to keep all the data of large enterprise applications in main memory. Compression also minimizes the amount of data that needs to be transferred between non-volatile storage and main memory on, for example, system startup. In SanssouciDB, data values are not stored directly, but rather as references to a compressed, sorted dictionary containing the distinct column values. These references are encoded and can even be further compressed to save space (see Sect. 4.3). This dictionary-compression technique offers excellent compression rates in an enterprise environment where many values, for example, country names, are repeated. Read performance is also improved, because many operations can be performed directly on the compressed data.

Write performance, however, is poor if new values need to be added to the dictionary. At some point, the complete column will need to be re-compressed. To overcome this problem we use a write-optimized data structure called the differential store as shown in Fig. 3.2. The differential store holds data that has been added to the database but has not yet been integrated with the data in the main store. It also only stores references to a dictionary but in this case the dictionary is unsorted to allow for fast updates. The differential store, together with the main store represents the current state of the database, therefore queries must be directed to both. Since the differential store usually requires more storage space per record than the main store, we do not want it to grow too large. To prevent this, we use a merge process to periodically move data from the write-optimized differential store into the main store. We will now briefly discuss some of the other key features of SanssouciDB.

When the query predicate has a high selectivity, that is, when only a small number of all possible rows are returned, scanning results in too much overhead even in a column store. For columns that are often queried with highly selective predicates, like primary or foreign key columns, SanssouciDB allows the specification of inverted indexes (Fig. 3.2 and Sect. 5.1).

In case of server crashes or power failures the primary data copy in an IMDB is lost. To allow for recovery in these circumstances, SanssouciDB writes a log and keeps snapshots of the database in non-volatile memory. These activities are tailored to the specific needs of an in-memory column-oriented database, as will be shown in Sect. 5.9.

Since SanssouciDB is tailored to business applications, it frequently needs to store and reconstruct business objects. These business objects are hierarchically structured and are mapped to the relational model to form a hierarchy of tables. To speed up operations on business objects, SanssouciDB provides the object data guide, which will be discussed in Sect. 5.3.

To reduce the need for locking and to allow us to maintain a history of all changes to the database, we adopt an insert-only approach (see Sect. 5.6). A consequence of this is that data volumes increase over time. Our aim is to always keep all relevant data in main memory, but as new data is added over time this becomes impossible. To ensure low latency access to the most recent data we make use of data aging algorithms to partition data into active data that is always kept in main memory, and passive data that may be moved to flash-based storage if necessary (see Fig. 3.2 and Sect. 5.2). The history store, which is kept in non-volatile storage, is responsible for keeping track of passive data. Keeping the history allows SanssouciDB to execute time-travel queries, which reflect the state of the database at a user-specified point in time.

To take full advantage of the performance improvements that SanssouciDB offers, new application development paradigms must be adopted. In particular, as much of the application logic as possible must be pushed down to the database (see Sect. 6.1). In Part III, we discuss this and other aspects of how a database architecture like that used in SanssouciDB can change the way enterprise applications are written and used.

3.4 Conclusion

We have described a novel architectural approach to the design of the database layer for enterprise applications, based on small clusters of high-end commodity blades with large main memories and a high number of cores per blade. SanssouciDB, our prototype database system, stores tables in compressed columns and keeps the primary data copy entirely in main memory. Non-volatile memory is only used for logging and recovery purposes, and to query historical data. Many requirements from business applications have influenced the design and functionality of SanssouciDB. These include the need to reduce locking as much as possible to allow for concurrent database access, the support for time-travel queries, and optimized business object management. Throughout the rest of the book, we offer insights as to why our approach is preferable for current and future enterprise applications.

Part II
SanssouciDB: A Single Source of Truth Through In-Memory

Chapter 4
The Technical Foundations of SanssouciDB

Abstract In this chapter we describe the detailed physical and logical foundations of SanssouciDB. Simply running an existing database on a machine with a lot of main memory and a large number of cores will not achieve the speed-up we are seeking. To achieve our goal of giving business users information at their fingertips, we need to create a database that fully leverages these technological advantages. Chief among these are the efficient use of main memory, and the parallelization of tasks to take advantage of all the available cores. In addition to these physical considerations, we also use logical design to increase performance. We lay out data in a way that suits the operations being carried out on it and also minimize the amount of data that needs to be processed.

In-memory and column-oriented DBMSs have existed for a number of years [2, 56], but their use in enterprise systems has been limited to analytical applications. Similarly, some current enterprise applications are optimized for parallel processing, but not the whole application stack. Recent hardware advances, specifically the reduction in the cost per megabyte of main memory, and the increase in the number of cores per CPU on commodity processors, have meant that the benefits of this new technology can now be applied across the enterprise application stack for the first time.

Understanding memory hierarchies and parallel processing are the main technical foundations that will enable developers and users to take advantage of the hardware developments mentioned above, but they are not enough on their own. Radically new designs of both the database and application layers are needed for SanssouciDB to fully exploit modern hardware, and thereby improve the performance of any enterprise applications using it.

H. Plattner and A. Zeier, *In-Memory Data Management*,   39
DOI: 10.1007/978-3-642-29575-1_4, © Springer-Verlag Berlin Heidelberg 2012

Fig. 4.1 Cost of data access with increasing stride

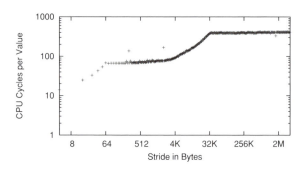

4.1 Understanding Memory Hierarchies

The goal of this section is to explain the concepts of main memory data processing. The first step is to describe the general structure of memory and the components of which the memory structure consists. We discuss its different components and the hierarchy level to which they belong and how they can be accessed (for example random or sequential). For optimizations we focus at first on latency penalties. In addition to understanding the general access patterns, we will examine main memory from a programmer's point of view.

4.1.1 Introduction to Main Memory

A program can be separated into two parts: the actual executed program code as instructions and all related data. The code is basically everything that controls the program flow and execution, while the data part holds all the variables and other program components. Before the CPU can process any data that is persisted on a non-volatile medium, such as flash disk it has to be loaded into main memory. Main memory is built from DRAM and should provide a constant access time to any memory location.

This is not a completely accurate description of the system's behavior. It is possible to validate the assumption of constant access time by a simple experiment: given a memory region spanning the available amount of main memory, a program now traverses this region by accessing a constant number of addresses but varies the distance between them. The assumption is that the average access time is independent of the actual distance between two addresses (stride) and the plotted result shows a straight line. The result looks differently. Figure 4.1 illustrates that depending on the stride, the experiment shows different access times per value. The obvious reason for this is that when data is loaded from main memory it is cached in different caches close to the CPU.

Fig. 4.2 Cost of data access
to an area of varying size

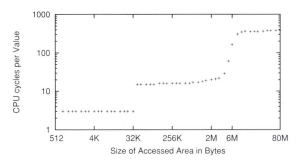

Using this experiment, it is possible to detect differences in block-sizes but not to detect the actual capacity of the cache. In another experiment, the stride between two items is kept constant, but the size of the memory region that is scanned is increased incrementally, see Fig. 4.2. This memory region is scanned multiple times such that if there is a cache, all data is already loaded. If the scanned memory region is smaller or equals the size of the cache no additional loading latency occurs. If the memory region is larger than cache, evictions occur and thus more load penalties become visible. This experiment highlights the different cache levels and allows the calculation of the size of each cache level.

Cache memories are subdivided into fixed-size logical cache lines, which are an atomic unit for accessing data. Caches are filled on a per-cache line basis rather than on a per-byte basis. To load a value from main memory it is necessary to iterate over all intermediate caches subsequently. Consequently, with stepping down in the memory hierarchy the CPU cycles needed to access certain values increases. For example, accessing main memory can consume up to 80 times the number of CPU cycles than an access to the CPU-near Level 1 cache consumes. This factor highlights the necessity of data locality in pure main memory persistent data management. Avoiding these so-called cache misses is one of the primary goals for optimizations in in-memory based databases. As in disk-based systems, the appropriate data management has a huge impact on the processing performance, regardless how the in-memory processing increases the application performance by leaving out the disk access. Furthermore, the cache line based pattern leads to an effect of reading too much data if data is accessed in a non-optimal way. For example, particularly in analytical workloads—which are mostly attribute-focused rather than entity-focused—the column-wise data representation is useful in a sense that only a small number of attributes in a table might be of interest for a particular query. This allows for a model where only the required columns have to be read while the rest of the table can be ignored. Due to this fact, the read cache line contains only the data needed to process the request, thus exhibiting a more cache-friendly memory access pattern . In case of row-oriented data storage, the cache line would be polluted with other attributes of the tuple depending on both, the width of the tuple and the size of the cache line. Attributes of a certain tuple can even be on separate cache lines leading to another cache miss while accessing two attributes adjoined to each

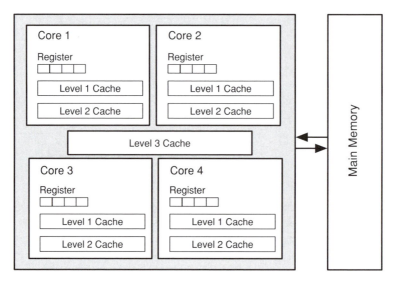

Fig. 4.3 Organization of CPU memory

other. Also, this effect can be observed when comparing disk-based row and column stores on page level. Another advantage of sequential memory access is achieved by the prefetching strategies of modern CPUs. This CPU-inherent technique allows prefetching of another cache line when reading and processing a cache line; that is, it speeds up the access to the proceeding cache line.

The hierarchy between main memory and a modern CPU usually includes the CPU registers, at least two levels of CPU caches, the Translation Lookaside Buffer (TLB) and the main memory itself, which in turn is organized into blocks. A more technical and detailed description of the purpose of the TLB can be found in [80]. Data that is about to be processed is transmitted from main memory towards the CPU cores through each of the memory layers. In an inclusive caching hierarchy,[1] which will be the focus of this section, every layer provides a cache for the underlying layer that decreases the latency for repetitive accesses to a piece of data. A request for a piece of data that is not currently stored in a cache is called a miss. A full miss, that is, a needed piece of data that is only present in main memory, results in an access to main memory and the transmission through all layers of memory. The time this takes is determined by two factors: the bandwidth and the latency of main memory.

A common setup of a multi-core CPU with Level 1, Level 2 and Level 3 cache is shown in Fig. 4.3. The bandwidth of a data transmission channel is the amount of data that can be transmitted through the channel in a given time period. It is usually measured in bytes or bits per second. When processing data from main memory it

---

[1] In an inclusive caching hierarchy the data cached on the lower levels is included in the higher level caches, while in an exclusive hierarchy a certain cache line can is stored exactly in one cache level.

Fig. 4.4 Memory hierarchy

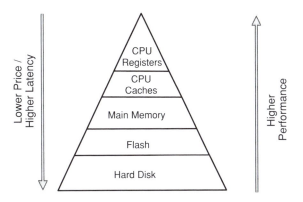

must be transmitted to the processing core through a number of layers. These include the following:

- From the main memory through the Memory Channel (for example QPI) to the Level 3 cache that is shared among all cores of the CPU.
- From the Level 3 cache through the CPU-internal bus to the Level 2 cache that is unique per CPU core.
- From the Level 2 cache through the core-bus to the Level 1 cache of the CPU core.
- From the Level 1 cache through the core-bus to the registers of the CPU.

The bandwidth of the channel from main memory to CPU core is the minimal bandwidth of any of the channels. Some channels like the QPI or the CPU-internal bus are shared among multiple processing units (cores), which might decrease the effective bandwidth of the channel. In the CPU architecture shown in Fig. 4.3 the Level 1 and Level 2 cache are private to the processing core and relatively small. The Level 3 cache is shared among all cores of the CPU and can be as large as 12 MB for example.

The latency of a storage is the time between the request for a piece of data and the beginning of its transmission. The latency of transistor-based memory is in its effect comparable to the seek time of a hard disk. The cause is very different: instead of the mechanical inertia of the disk arm it is the time to decode a memory address and connect the transistors that contain the requested piece of data to the bus. For caching memories, the latency also originates from the time it takes to determine if, and if so where in the memory, a given block is cached.

With IMDB it is crucial to optimize for the different properties of the memory hierarchy and all data access through all levels, including reducing the number of CPU stalls while waiting for memory. The overall goal is an optimal memory layout of all stored data with regards to the given workload in terms of access patterns. For all optimizations it is important to keep the memory hierarchy pyramid of Fig. 4.4 in mind. The closer a certain memory is located to the CPU the smaller and faster the memory is. In this case performance and size correlate with price. Making the close to the CPU caches larger requires more silicon due to the implementation details of Static Random Access Memory (SRAM) thus raising the price for a CPU.

4.1.2 Organization of the Memory Hierarchy

A successful cache organization and cache replacement strategy tries to maximize the number of cache hits over the number of cache misses. In order to achieve a high hit/miss ratio the fundamental principle behind the memory hierarchy has to be considered. This principle is called locality of reference. There are two different kinds of locality regarding memory access: spatial locality and temporal locality. Spatial locality refers to the fact that the CPU often accesses adjacent memory cells whereas temporal locality describes the observation that whenever the CPU addresses an item in memory it is likely that this item will be accessed again in the near future. A cache takes advantage of spatial locality by loading memory blocks including the requested item, as well as some of its neighbors into the cache.

Whenever the CPU requests a certain memory item the cache is searched for the corresponding cache line. In cases when a cache miss occurs, this search causes additional overhead for the entire memory access. Thus, to minimize this overhead a memory block can only be placed in special cache lines. This organizational concept is referred to as cache associativity. One can distinguish among three different types of cache associativity:

1. *Direct mapped*: a particular memory block can only be mapped to exactly one particular cache line. Memory blocks that map to the same cache line can never reside in the cache at the same time. A cache miss caused by colliding memory blocks is called a conflict miss.
2. *N-way associative*: in this case, each memory block maps to a distinct set of N cache lines. The position of the block within its associated set can be freely chosen.
3. *Fully associative*: in a fully associative cache, a memory block can be stored at any position.

A higher level of associativity allows a more flexible replacement strategy, that is, a more flexible choice of those cache lines that have to be removed in cases of loading new data from memory to a full cache. The cache lookup time increases because in the case of a fully associative cache, all cache lines have to be scanned in the worst case if the CPU requests an address from memory. While a higher level of associativity might increase the cache-hit rate, the latency of the cache increases. In practice, the level of associativity is chosen to be less or equal to 8-way associative, which has turned out to be a reasonable compromise between cache latency and hit rate.

In modern CPUs, in addition to caches located directly on the CPU core, an uncore cache is added that is shared among all cores on the die. As a result, this approach adds another level of indirection to the memory hierarchy but allows further cache aware optimizations for multi-core software.

Temporal locality of memory references can be exploited by the cache replacement policy. It describes the strategy by which old cache lines are replaced when loading new data from memory to the cache. Examples of viable replacement rules are First In, First Out (FIFO) and Last In, First Out (LIFO). None of these rules assumes that a recently used cache line is likely to be used in the near future. Thus, an optimal cache replacement strategy (also described as Belady's optimal page replacement policy) would replace the cache line that has not been used for the longest time—this strategy is called Least Recently Used (LRU).

4.1.3 Trends in Memory Hierarchies

In the last few years the performance gap in terms of latency between storage media and the rest of a computing system has continued to widen rapidly. To overcome this obstacle and to maintain performance growth rates, much research effort has been put into developing new storage technologies that overcome the limitations of traditional magnetic disk storage. Still under development is a group of new and promising storage technologies known as storage-class memory. Common properties of this new type of storage media are non-volatility, very low latencies, solid-state implementation with no rotating mechanical parts, inexpensiveness, and power efficiency.

An early version of storage-class memory technology that tried to overcome the limitations of traditional hard disks is flash memory. In flash memory, data is stored in cells consisting of floating gate transistors that are able to keep their current even when the power supply is switched off. No current refresh is necessary, which reduces the power consumption vis-à-vis volatile DRAM. Additionally, flash memory provides fast random read access times that are significantly lower than the access times of hard disks [87]. Flash memory has some drawbacks. A flash memory cell cannot be updated in-place, meaning that it has to be erased before a new value can be stored in the cell. Furthermore, the time to inject charge into a cell is longer than the time needed to read the current cell status. The write performance of flash memory depends on the access pattern and decreases significantly for random writes compared to sequential writes. This asymmetric read/write performance has to be considered in the design of flash-aware algorithms in order to leverage the potential of this storage technology [104]. A major disadvantage of flash memory versus disk storage is its limited physical durability. Each write access slightly damages a memory cell decreasing the lifetime of flash memory (about 104–105 writes) significantly compared to main memory (about 1,015 writes). To alleviate the limited duration of flash memory a technique called wear leveling is used that tries to equally distribute write accesses over the existing physical address space to avoid write hot spots.

In the context of traditional database management systems, the usage of flash memory has been examined for two different scenarios. First, flash technology can be used like memory as an additional cache hierarchy level. The problem with this scenario is the limited durability of flash memory contradicting the extensive read and write access of a cache. It has been often proposed that flash memory acts as a persistent storage medium. In this case, the data structures and access routines of the DBMS have to be adapted in order to gain the full performance potential of flash memory. In particular, the asymmetry between read and write access has to be taken into account ([104] describes an in-place logging approach exploiting the properties of flash storage).

In the scope of in-memory database management systems, the role of flash and other storage-class memory technologies could be twofold. First, flash volumes can be used as major persistent storage devices, leaving disks as backup and archiving devices. The insert-only paradigm of an in-memory database matches the advantages of flash memory. In an insert-only database the number of random writes can be reduced if not eliminated at all and the disadvantage of limited durability is alleviated by the fact that no in-place updates occur and no data is deleted. Second, the low readout latency of flash storage guarantees a fast system recovery in the event of a system shutdown or even failure. In a second scenario, flash could be used as memory-mapped storage to keep less frequently used data or large binary objects that are mainly used during read accesses. The DBMS can transfer infrequently used columns to a special memory region representing a flash volume based on a simple heuristic or manually by the user. The amount of main memory can be reduced in favor of less power consuming flash memory, thus reducing the overall energy consumption of the system while maintaining performance.

Another promising storage-class memory technology that has been developed recently is Phase Change Memory (PCM). In phase change memory chalcogenide alloys are used as bi-stable storage material. These materials exist in two stable phases: the first is amorphous and provides high electrical resistance whereas the second one is polycrystalline and provides low electrical resistance. PCM has the following major advantages over flash: first, PCM can be updated in place. Unlike flash memory, it does not require a cell to be erased before a new value can be written to the cell. Thus, the performance for random writes is better than for flash memory. Second, the write endurance of PCM is several orders of magnitude higher than for flash memory. In the scope of in-memory databases storage-class memory with its improved physical endurance properties and increased write performance will likely be able to replace disk and become the persistent storage medium of choice closing the gap between main memory and flash even more.

4.1.4 Memory from a Programmer's Point of View

The following section explains how to best access memory from a programmer's point of view. The goal is to obtain a better understanding of how memory access is triggered from a program and what the direct consequences of this memory access are. In addition, we explore potential optimization possibilities when working with main memory.

Allocating Memory and Writing it Back

When programmers deal with objects or other data structures, they are often not aware of the fact that each access or modification operation to such a data structure directly interacts with main memory. In higher level programming languages memory access is often hidden from the programmer and maintained by a virtual machine or any other external memory manager. In this section, we will focus on memory access in a low-level programming language like C or C++. As described earlier, every access to main memory follows a strictly block-oriented approach. It is, therefore, very important to take care of the sizes of the different data structures and their alignment.

With each instruction the programmer can load different memory areas without interfering with another level of the memory hierarchy. At most 128 bits can be loaded into a single CPU register and working with the content of this register does not require any additional loading from the L1 cache and thus does not incur an additional loading latency. When accessing the contents of a single cache line—depending on the CPU architecture—64 bytes can be accessed without incurring an additional loading latency. For example, this means that if this cache line is already loaded in the L1 cache, the access to any word stored in the cache line only requires one cycle to be loaded to the CPU, versus five to ten cycles if the cache line has to be loaded from the L2 cache. In newer CPUs with core-shared Level 3 caches this number increases even more.

With respect to these loading latencies it becomes crucial to minimize the concurrent access to multiple cache lines and different blocks along the memory hierarchy.

A common optimization technique to increase the performance of accessing main memory and to create a cache-conscious design is to use only aligned memory for larger memory areas. Alignment defines that the newly allocated memory region points to an address that is a multiple of a given alignment value. As a result, it is possible to allocate memory blocks that are aligned to the width of a cache line, meaning the beginning of this memory region always points to the beginning of a cache line without triggering skewed data access. On POSIX compatible operating systems this can be achieved by using the function posix_memalign() to allocate a memory region aligned to a variable multiple of a cache line size.

Another common optimization method is called padding, which is an approach to modify memory structures so that they show better memory access behavior but requires the trade-off of having additional memory consumption. One example is a method typically applied to complex data structures when using the C or C++ programming languages. If the size of the structure is not a multiple of 4 bytes, an additional padding is introduced to add the difference and thus avoids unaligned memory access. In the context of main memory databases one use for padding is when storing wider rows. If the width of the tuple is not a multiple of a cache line, the width of the tuple is padded so that its width becomes a multiple of a cache line size. Again this padding avoids unaligned memory access, and if this is the case, no additional cache misses are caused when reading the tuple, although the read width is strictly smaller than a cache line.

Listing 4.1 (Example for Structure Padding)

```
#include <stdio.h>

typedef struct
{
unsigned first;
unsigned second;
char third;
} val_t;

int main(int argc, char* argv)
{
printf("%ld\n", sizeof(val_t));
return 0;
}
```

The example in Listing 4.1 shows the definition of a `struct` in C. Given only this example what would one expect to be the size of this `struct`? We assume that unsigned values use 4 bytes and a char is stored as a single byte value. From a logical perspective the size of the `struct` would be 9 bytes: two 4-byte integers plus the single byte value. For the CPU accessing 9 bytes is not an easy task because sub-word access—assuming a CPU word size of 4 bytes—triggers additional instructions to extract the values from the given word. Thus, the compiler will automatically pad the `struct` to a size of 12 bytes so that additional instructions are avoided. In addition to narrow `struct` padding this example shows the advantages of alignment and wide padding. Assuming a cache line size of 64 bytes, five `val_t` values will fit on a single cache line. If all values are stored contiguously, the access to the 6th will incur two cache misses since it is stored across two cache lines. If we want to guarantee that each access to a `val_t` value will at maximum generate one cache miss, we have to add additional padding. As a result for every fifth `val_t` additional padding of 4 bytes are added to fill the space until the beginning of the next cache line. This optimization is useful where only a certain subset of the data is retrieved. If all data is read from the memory region, the prefetcher of the CPU will load the adjacent cache lines and thus hide the additional penalty. If only a fraction of the data is retrieved and in addition only those tuples that are stored across two cache lines are accessed the penalty is higher. Depending on the expected workload it can be advisable to use or not to use this optimization.

The last performance optimization technique we present in this section is called blocking or pointer elimination. The goal is to store as many fixed-size elements consecutively in memory and address them indirectly via offsets. The size of the block should be a multiple of a cache line. First, the advantage is that to link subsequent elements no explicit next pointer is required. This optimization saves memory and allows more elements to fit into one block. Second, this technique improves the sequential reading performance. This section showed only a small set of possible optimizations with respect to cache-conscious data access, but there are many more for the various use cases, including cache-conscious trees and hash maps [145]. They all share the common goal of exploiting the block level structure of the memory hierarchy and favor sequential reading through main memory. In addition to such cache-conscious optimizations, algorithms can also be optimized in a cache-oblivious way. An algorithm is cache-oblivious if no program variables dependent on hardware configuration parameters, such as cache size and cache line length need to be tuned to minimize the number of cache misses and the algorithm still exploits the behavior of the memory hierarchy implicitly.

Sequential Reading

As explained in the previous paragraph, it is possible to deduct that certain access patterns are better than others. As in other layers of the memory hierarchy, the most important access pattern is sequential access to data stored in main memory. Referring back to Fig. 4.1, we can basically show the difference between random and sequential access. In Fig. 4.1, the sequential access is on the left-most part of the graph. Accessing multiple values from one cache line that was previously cached takes only a few cycles. Accessing the same values from a different memory location can take up to 200 or even more CPU cycles.

From a programmer's point of view, the ultimate goal is to align all data-intensive access patterns so that they read the data sequentially and avoid random lookups. In modern database systems storing the data in a column-wise fashion and compressing the values of each attribute can improve the sequential access to data. As a result, the virtual reading bandwidth can be increased especially for sequential access operations.

There are access patterns where random reads are necessary, such as in join operators. In those cases the goal is to carefully observe the memory access sequence and used algorithms. In many cases it is then possible to create blocks during the evaluation to decrease the amount of random memory accesses. A common example is the CSB+ tree [145]. Typically, during search a operation using the tree, every access to a single key will trigger a random access and if the size of the tree is large enough a cache miss due to the distribution of the keys in the tree. The CSB+ tree is used in different components in SanssouciDB, such as in the dictionary of the differential store to keep a sorted list of all newly inserted values. In comparison to the traditional B+ tree the keys on each node are stored sequentially in memory to eliminate pointers to find the next key. The size of the node lists is aligned to the size of the cache line to avoid unnecessary cache misses. As a result the search performance of a CSB+ tree is better than the B+ tree due to its better memory structure.

The Impact of Prefetching

One of the obvious problems with loading memory is that loading data incurs additional loading latencies and the executing program stalls for this amount of time.

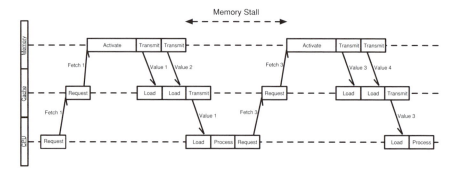

Fig. 4.5   No prefetching

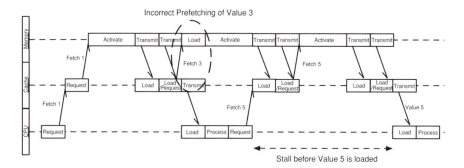

Fig. 4.6   Incorrect prefetching

To avoid this, CPU manufacturers implement prefetching algorithms that will load co-located data as soon as the first load is finished. If the program execution takes longer than reading the next cache line, the access to this second cache line will incur no additional stall.

The behavior of prefetching is shown in the above activity diagrams. Figure 4.5 shows an execution diagram with no prefetching enabled. The access to the main memory is handled separately for each cache line, so that a cache miss always occurs after the previous cache line has been read and the CPU has to wait until the load from the main memory is finished. With prefetching turned on—depending on the prefetching strategy—the adjacent cache line is loaded as soon as the loading of the first one is finished. As a result, the CPU can continue reading the memory region as soon as it finished reading the first cache line. Typically no additional memory stalls occur. Figure 4.6 shows an incorrect behavior of prefetching. As soon as the request for the first cache line is finished the next adjacent cache line is loaded. In this case only the third cache line is actually read, leaving the second untouched. When the CPU wants to access values from the third cache line and the memory bus is still blocked by reading the second cache line, memory stalls occur. Now the

CPU has to wait until the second load is finished. If this pattern occurs repeatedly it can induce a severe performance impact. In current CPUs different prefetching strategies are implemented. Typically, the CPU is able to detect if it is advisable to use the prefetcher. However, it is still possible to gain a few percent performance win when the strategy is chosen manually.

Performance Measuring for Main Memory Systems

In addition to understanding how the memory architecture works and what defines the parameters that influence software systems it is important to measure the different properties of such software systems. To achieve the best possible performance for any software system it is important to carefully observe the different performance properties of this system. In contrast to classical I/O devices such as disks, for main memory systems it is possible to perform performance measuring on a very fine granular level. Modern CPUs provide special registers to count CPU-related performance events like CPU cycles, cache accesses or executed instructions. Based on those observations it is possible to tune the behavior of data structures or algorithms.

Virtualized Access to Main Memory

Virtualization provides the necessary means to make main memory systems interesting even for smaller applications. The general assumption is that virtualization will incur a big performance penalty for in-memory systems. The goal of this section is to show the impact of virtualization on IMDB performance. While at first thought one would expect significant cache trashing due to the fact that the design of modern IMDB often directly exploits specific concepts of the underlying hardware (for example, prefetching, vectorization [185]), we show that virtualization does not in fact inhibit these optimizations. The reason is that IMDB essentially eliminate one layer in the storage hierarchy since data that is being operated upon is kept in DRAM. When solely reading memory pages the hypervisor is often not required to be called, which limits the overhead incurred by virtualization.

To show the impact of access to main memory in virtualized systems we conducted two experiments. The goal of the first experiment determines the size and access latency of the different CPU caches both in a virtualized and non-virtualized environment. In this experiment the MonetDB Calibrator, as shown in [110], is used to determine these properties.

Table 4.1 shows the results of a measurement run on both the physical and virtual test systems. In both cases we used a data region of 1 GB. From the result of the test run we can derive an interesting fact: in both cases the detected sizes for L1 and L2 cache are identical, as are the times for the latency. The only observed difference between both systems is the cost for TLB misses. While on the physical system no TLB—even though the TLB exists—could be detected, the calibrator detects a TLB on the virtualized system. The reason for this behavior is that the latency of

Table 4.1 Comparing access latencies between physical and virtualized system

Description	Physical system	Virtualized system
L1 size	32 kB	32 kB
L1 miss latency	10 cycles	10 cycles
L2 size	6 MB	6 MB
L2 latency	197 cycles	196 cycles
TLB miss latency	–	23 cycles

the TLB is hidden behind the latency of an L2 cache miss. In a virtualized system, address translation is handed down to the hypervisor and creates additional overhead. Although this micro-benchmark shows that the absolute memory access performance in a virtualized system is equivalent to the physical system, the additional overhead for address translation can become a problem, since it becomes more important when more data is read. In the second series of experiments we wanted to evaluate the performance impact for main memory systems [67]. The setup presents two experiments specific to the main memory storage engine:

- *Projectivity*: in this experiment we load two tables: one with 100 single column groups and another table with one group of 100 columns. During the experiment we increase the number of projected attributes. The goal of this benchmark is to observe the impact of partial projections in different setups. The experiment shows a similar result compared to the non-virtualized execution of the query with the additional overhead of TLB misses.
- *Selectivity*: in this experiment we load two tables: one with 16 single column groups and one table with one group spanning 16 attributes. We now execute a query that reads the complete width of the table while varying selectivity from 0 to 1 using a list of randomly uniform distributed positions is generated at which the data shall be read.

The general observation shows that the virtualized system behaves like the physical system. Furthermore, it is possible to deduct that the performance of the virtualized system depends on the amount of data read. This assumption can be verified when looking at the results of the selectivity experiment (Figs. 4.7 and 4.8). For each column a distinct memory location has to be read, requiring loading a complete cache line although only a fraction of it is used. On the contrary, when reading selective data from the row store only the requested cache line is touched and read completely. On the physical system the latency of the TLB is hidden by the L2 miss and cannot be measured, while the address translation on the virtual system is more expensive and the latency of the TLB becomes measurable. The complete series of experiments with a more detailed context is presented in [69]. The conclusion of the above experiments is that virtualizing IMDB is possible and feasible.

Fig. 4.7 Results from the selectivity experiment for column storage

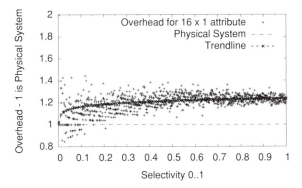

Fig. 4.8 Results for the selectivity experiment for row storage

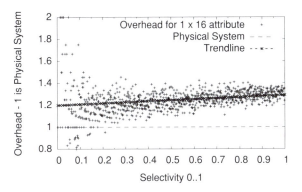

4.2 Parallel Data Processing Using Multi-Core and Across Servers

Enterprise database systems grow in multiple dimensions during their lifetime, for example, in database size and number of users. In the past, increasing processor clock speeds have meant that simply buying newer, faster hardware has been enough to cope with the growing workload, but since 2001 a paradigm shift in CPU development has taken place—away from increasing the clock rate toward increasing the number of cores per CPU [80, 88]. Designing systems to take advantage of this new trend toward parallel data processing is a prerequisite for success in the future. This is not always straightforward, however, and requires a new approach to designing systems; in particular, operations that require sequential processing, such as locking, should be avoided where possible [78].

The first part of this section introduces the basic concepts of scalability and parallel data processing in database systems. Finally, we introduce the specific concepts around parallel data processing and scalability developed for SanssouciDB.

4.2.1 Increasing Capacity by Adding Resources

Scalability is the capability of a system to yield an increase in service capacity by adding resources. A scalable application should only need more resources for scaling and should not require significant modifications of the application itself [115].

Ideally, a system should be linearly scalable, which means that doubling hardware resources should also lead to a doubling of service capacity, for example, doubling the throughput of a system. To measure scalability, it is a common approach to quantify the speed-up of a sequential system in comparison to a parallel system [47]. The speed-up of a system can be defined as follows

$$speedup_1 = \frac{time(1, x)}{time(n, x)} \tag{4.1}$$

where $time(n, x)$ is the time required by a system with n processors to perform a task of size x.

According to Amdahl's Law, the maximum speed-up of a system can also be defined in terms of the fraction of code that can be parallelized [8]

$$speedup_2 = \frac{1}{r_s + \frac{r_p}{n}} \tag{4.2}$$

where r_s represents the sequential fraction of the program, r_p the fraction of the program that can be parallelized, and $r_s + r_p = 1$. Parameter n represents the number of processors or cores. Assuming that the number of processors is unlimited, the speed-up can then be calculated as

$$speedup_3 = \frac{1}{r_s} = \frac{1}{1 - r_p} \tag{4.3}$$

The graph in Fig. 4.9 shows what theoretical speed-up can be achieved assuming an unlimited number of processors.

In the figure, the fraction of parallelizable code is the amount of program code that can be executed in parallel by an infinite number of processes or threads. A system that scales in an ideal way has two key properties: a linear speed-up and a linear scale-up. Linear speed-up means that doubling the hardware halves the time a task takes to complete, whereas linear scale-up means that doubling the hardware doubles the task size that can be processed. In addition to Amdahl's law, there are three technical barriers [45]:

- *Startup*: the time needed to start a parallel operation. If thousands of processes must be started, this can easily dominate the actual computation time.
- *Contention*: the slowdown each new process imposes on all others when accessing shared resources.

Fig. 4.9 Maximum speed-up for fraction of parallelized code

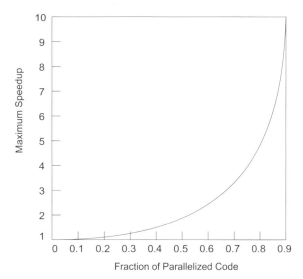

- *Skew*: the deviation in resource utilization and processing time among parallelized steps. As the number of parallel steps for a task increases, the average size of each step decreases, but it is still possible for one or a few steps in the process to take a much longer for completion than the mean time. The total execution is thus skewed by these few long running tasks.

The time required to process one task containing several steps that can be executed in parallel is then the processing time of the slowest step in the task. In cases where there is a large amount of skew, increased parallelism improves elapsed time only slightly.

To cope with an increasing load on a DBMS, a database server can be equipped with more main memory, more non-volatile storage, and it can use faster CPUs. This is often referred to as scale-up [45] or vertical scaling.

Beyond a certain point, increasing the performance of a single database server gets prohibitively expensive [80]. The database system can also be scaled horizontally on multiple machines, often referred to as scale-out. Due to the fact that network bandwidth has increased enormously within the last decades, horizontal scaling of a database system is often a more cost effective way to meet the demands of today's database applications and their users, than scaling up a single machine [163]. Using more than one server is also important as a way of providing high availability (see Sect. 5.8). In the following sections we will discuss how scale-up and scale-out can be achieved using different parallel system architectures and how SanssouciDB makes use of them.

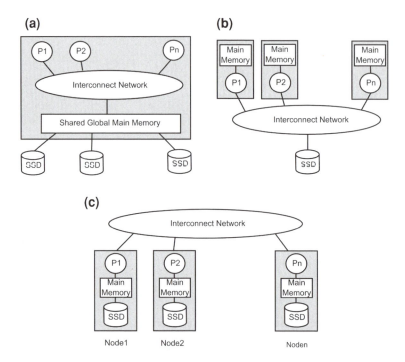

Fig. 4.10 Parallel architectures: **a** shared-memory, **b** shared-disk and, **c** shared-nothing

4.2.2 Parallel System Architectures

Hardware architectures for database systems can be divided into the following three broad categories based on the hardware components that are shared among the processors in the system:

- *Shared memory*: all processors share direct access to a common global main memory but have indirect access the non-volatile storage (Fig. 4.10a).
- *Shared disk*: all processors share direct access to the non-volatile storage, but each processor has its own private main memory (Fig. 4.10b).
- *Shared nothing*: each processor has its own memory and non-volatile storage and can act independently of the other processors in the system (Fig. 4.10c).

Shared-memory and shared-disk systems generally characterize the scale-up approach while shared-nothing is applied to scale out the database.

Shared Memory

In a shared-memory system, all processors access a shared global memory via an interconnection network. All data required by each of the processors must pass through this network. It is clear that this could become a major bottleneck in the system. To provide good performance, the interconnection network should have sufficient bandwidth and low enough latency to process data from all processors and disks at once, otherwise contention can have a significant impact.

To help reduce traffic through this interconnection network and to minimize contention, each processor has a large private cache, and the system can use an affinity scheduling mechanism [150] to give each process an affinity to a particular processor. This can, however, create skew and load balancing problems. Indeed, measurements of shared-memory multiprocessors running database workloads show that loading and flushing these caches considerably degrades processor performance. As parallelism increases, the contention problems increase, thereby limiting performance.

Shared Disk

In the shared-disk approach, contention is reduced by allocating a portion of the system's main memory to each processor, while still providing direct access to all disks. The database loads data from the shared disk into each processor's local memory where it can be processed independently of other processes. The major disadvantage of this approach is that if the data required by a processor cannot be found in its local memory, it must be fetched from disk, which exacts a high performance penalty.

Shared Nothing

The shared-nothing architecture minimizes contention by minimizing resource sharing. The key benefit in a shared-nothing approach is that while the other architectures move large volumes of data through the interconnection network, the shared-nothing design moves only queries and results. Memory accesses are performed locally, and only the result sets are passed to the client application. This allows a more scalable design by minimizing traffic over the interconnection network. Recently, shared-nothing database systems have become more popular as the availability of fast, inexpensive commodity components has increased.

Whether to scale up or scale out a system is often a question of the use-case. Services such as those provided by Google or Facebook typically use large clusters containing commodity hardware and focus on scaling out rather than scaling up. However, having lots of small commodity servers can lead to synchronization and consistency problems. Therefore, enterprise systems are typically scaled up before being scaled out and use high-end hardware that is more reliable and provides better performance, but at a higher cost.

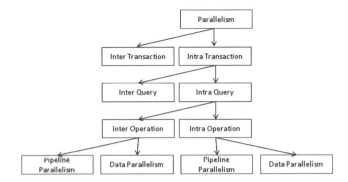

Fig. 4.11   Parallelization techniques at different granularity levels [144]

4.2.3 Parallelization in Databases for Enterprise Applications

Parallelization can be achieved at a number of levels in the application stack of an enterprise system—from within the application running on an application server to query execution in the database system. As an example of application-level parallelism, let us assume the following: if, at a certain point in time, an application needs to execute multiple independent queries, these queries can be sent in parallel to the database system using multiple connections. Obviously, this kind of parallelism requires manual action from the application programmer: they have to analyze and implement the application in such a way that independent queries can run in parallel.

In contrast, a salient feature of database systems is automatic parallelization. Relational queries consist of uniform operations applied to uniform streams of data. Each operator produces a new relation, so the operators can be composed into highly parallel dataflow graphs. We can classify the parallelization techniques as shown in Fig. 4.11.

Usually, multiple applications access the same data. Therefore, inter transaction parallelism, that is, simultaneously processing multiple transactions (from multiple users or application programs) is essential for all database systems. To achieve parallelism within a transaction (intra transaction parallelism), the application programmer can decide to issue multiple queries at once (inter query parallelism) as sketched above. Parallelizing a single query (intra query parallelism) is in the realm of the database system and out of the application programmer's influence. Database systems can implement inter operation parallelism and/or intra operation parallelism to achieve parallel query execution. In the first approach, the DBMS analyzes the query to identify operations that are independent of each other, for example, scans of different relations. These operations can be run in parallel. In the second approach, query processing algorithms can be parallelized internally. Obviously, the two approaches can be combined.

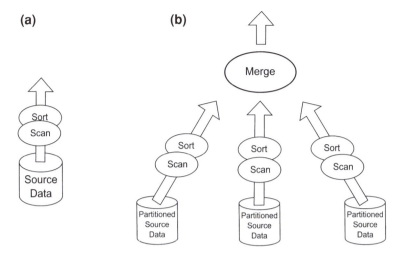

Fig. 4.12 **a** Pipeline and **b** data parallelism

Implementations for both inter-operation parallelism and intra-operation parallelism are based on two fundamental techniques: pipeline parallelism and data parallelism, as shown in Fig. 4.12.

By streaming the output of one operator into the input of a second, the second operator can begin working before the first operation is completed. In Fig. 4.12a, the sort operation can start sorting the output of the scan while the scan operation is still working. This is an example of pipeline parallelism. In a RDBMS, the benefits of pipeline parallelism are limited. First, relational pipelines are rarely very long. Second, relational operators such as aggregate and sort, both important for enterprise applications, do not emit their first output until they have consumed all their inputs. These operators can therefore not be pipelined. Finally, the execution cost of one operator is often much greater than that of the others which is an example of skew. In such cases, the speed-up obtained by pipelining is very limited.

By partitioning the input data among multiple processors, the execution of an operator can be split into many independent executions each working on a subset of the data in parallel. This partitioned data and execution model is called data parallelism (Fig. 4.12b) and can deliver significant performance improvements. The way execution threads are assigned is related to the way the data has been partitioned (see Sect. 4.4). This can be done in the following ways:

- *Position ranges*: the first thread operates on rows 0 to 999, the second thread operates on rows 1,000 to 1,999, and so on.
- *Value ranges*: the first thread operates on all values a–d, the second thread operates on all values e–h, and so on.
- *Hashing*: the first thread operates on all values whose binary representation of the hashed values starts with 00, the second thread operates on 01, and so on.

- *Round robin*: assume we have ten threads. The first element is assigned to the first thread, the second data element to the second thread, and so on, until the eleventh data element, which is then again assigned to the first thread, the twelfth to the second thread, and so on.

Aggregation, a commonly used operation in enterprise applications, can benefit from data parallelism. Each computation node in control of a partition can aggregate its own data and return results to a central merger node or to a set of merger nodes asynchronously. No node needs to wait for results from any of the other nodes. This is discussed in detail in Sect. 5.6.

4.2.4 Parallel Data Processing in SanssouciDB

SanssouciDB leverages parallelism along two dimensions; across blades and cores. Because its target hardware is a cluster of blades, SanssouciDB allows us to partition the data across multiple blades according to the partitioning schemes introduced in the previous section. To scale out efficiently with the number of blades, the query processing algorithms operating on the distributed partitions have to be designed in such a way that the operations on the partitions run, as far as possible, independently from each other.

The second dimension of parallelism SanssouciDB is designed for, targets the operations inside one blade, that is, multi-core parallelism with access to shared memory. Optimizing for shared-memory access and the best possible cache locality is the primary goal of these algorithms in an in-memory system like SanssouciDB. We will introduce the parallelization concepts below and give a more detailed discussion for parallel aggregation and join operations in Sect. 5.6.

SanssouciDB can alleviate the problems associated with parallel data processing that we discussed in the first section:

- *Startup*: the large number of cores available in a SanssouciDB system means we can limit ourselves to one process per core. Starting one process per core adds no significant overhead.
- *Contention*: the multi-channel memory controllers in the modern processors that we use in SanssouciDB reduce this problem for cores on the same machine [116].
- *Skew*: it is still possible for some parts of an operation's execution to dominate the total execution time; however, each part is executed so quickly that the overall impact should not be significant.

SanssouciDB also takes advantage of SIMD instructions to provide extra data parallelism [185]. The amount of possible parallelism depends on the number of objects that can be loaded into a register. Assuming a 128-bit register, the Streaming SIMD Extensions (SSE) instructions can work on up to four integers at a time. Figure 4.13 shows how the Parallel Add (PADD) instruction can perform four additions in a single clock cycle. It is important to note that SIMD is orthogonal to multi-core parallelism. Each core in a system can make use of the SIMD functionality independently.

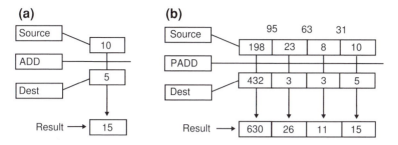

Fig. 4.13 **a** Scalar add and **b** parallel add

The column-oriented approach adopted in SanssouciDB is particularly well suited to SIMD execution. Data is stored in a structure of arrays, one array for each column in a given table. The arrangement means that all the data in a column is stored sequentially in main memory, making it easy for the SIMD instructions to pick four values at a time during a column operation.

Intra- and Inter-Operator Parallelism

Parallelizing query processing algorithms (intra-operator parallelism) is a prerequisite for database systems to scale with the number of available cores on a system and with the number of available machines in a cluster. Generally, we distinguish between algorithms tailored to shared-memory/shared-disk systems and those for shared-nothing architectures. The first group can utilize the shared memory/disk to operate on shared data, for example, to share intermediate results. The latter group explicitly needs to send data between nodes of the shared-nothing architecture.

SanssouciDB provides algorithms for both shared-memory and shared-nothing architectures. Since the disk is only required for persistence and not for data processing, shared-disk algorithms are not used. For now, we consider algorithms for shared-memory architectures.

On a shared-memory system, all execution threads can access the same data. This simplifies access to the input relation(s) and to intermediate results. There are various ways to partition the input data. The position ranges partitioning scheme imposes the smallest overhead, so SanssouciDB always uses this scheme to partition the input table.

The following is a list of algorithms that have been parallelized in SanssouciDB:

- *Scan*: a scan evaluates a simple predicate on a column. The result consists of a list of positions that indicate where the predicate matched. More complex predicates that involve multiple columns can then be evaluated based on the position lists returned from the scans. Parallelizing the scan is quite simple; each thread receives a certain partition of the column on which to operate. The results delivered by all threads are concatenated.

- *Aggregation*: this is typically implemented with the help of a hash table. The hash keys are composed of the combined values from the columns to group by. The hash values are computed based on the columns to be aggregated. Special hash table implementations allow concurrent insertions (without the need to lock the entire hash table). In SanssouciDB, multiple threads can fill this hash table in parallel. Section 5.6 gives a complete overview of aggregation algorithms, their interaction with hash tables and the specifics regarding SanssouciDB.
- *Join*: equijoins (see Sect. 5.6) can also be implemented with the help of hash tables. For both relations to be joined, a separate hash table is initialized. The hash key consists of the concatenated values from the columns to be joined. The hash value is a list of rows, where a particular join combination occurs. As in the parallel aggregation variant, both hash tables can be filled by multiple threads in parallel. The generation of the join result simply connects the row lists for the same hash value.

In summary, intra-operator parallelism on shared-memory/shared-nothing machines can be achieved by parallelizing query processing algorithms using multiple threads that operate on the same input data.

Note that parallel algorithms have been specifically developed for shared-nothing architectures. However, there is also a simple yet practical approach to shared-nothing algorithms, which exploits inter-operator parallelism: the algorithm to be parallelized is decomposed into various sub-algorithms that can run in parallel on the partitions stored at the various nodes of the shared-nothing system.

Operator trees are data structures commonly used to internally represent a query in a database system. In such a structure, the nodes are operations and the edges indicate data flow. Often, the internal representation directly reflects the relational algebra representation of a query (see Sect. 5.1). As in the evaluation of a relational algebra expression, the query is evaluated from bottom to top and the root node returns the result. Often, a query tree contains independent branches. These branches can be evaluated in parallel, providing inter-operator parallelism.

Assume we have two tables Sales and Forecast. Both tables have columns Product (P) and Location (L), as well as a key Sales (S) or Forecast (F), which contain entries that represent the sales forecast data for a certain product and location. Assume we are interested in analyzing which product at which location had lower than forecasted overall sales in 2010. The query plan in SanssouciDB is shown in Fig. 4.14.

The plan can be interpreted as follows: to restrict the data volume to be processed, a scan returns only the rows for 2010 of columns P, L, S or P, L, F (depending on the table). The group-by operations group the resulting rows by P and L and sum column S on Sales and F on Forecast. The result is joined (again on P and L) by the subsequent join and columns S and F are passed on, as specified by the projection. Finally, only those rows that have a smaller aggregated sales value than forecasted are returned. Because the operations are executed on different input tables, they are independent of each other. The left branch below the join operation in Fig. 4.14 can be executed in parallel to the right branch (inter-operator parallelism). Note that inter-operator

Fig. 4.14 Query execution plan for the sales analysis query

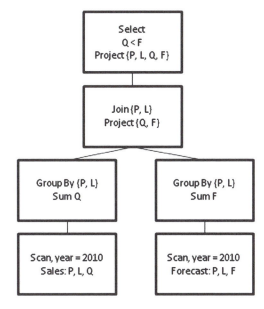

parallelism is orthogonal to intra-operator parallelism: all operators in Fig. 4.14 can utilize multiple threads internally to further increase the degree of parallelism.

Because SanssouciDB runs on a shared-nothing/shared-memory architecture, the query-plan generator and the query execution engine have to be able to generate and evaluate distributed queries. This is accomplished by adding a location specifier to the plan operations. As an example, assume, that the Sales and the Forecast tables are horizontally partitioned based on the values of P and L (that is, there are no rows with the same value of P and L stored in two different partitions). The partitions of one table are stored across two nodes. Thus, Sales is stored at Node1 and Node2, while Forecast is stored Node3 and Node4. The plan generated for this setting is shown in Fig. 4.15.

The structure of the query plan is quite similar to the previous one: the scan is followed by an aggregation and then by a join. Because the data is partitioned by P and L, the scan operation can be executed on the joined partitions separately. The final operation is a merge. Again, the independent operations can run in parallel giving us inter-operator parallelism. Because the data is partitioned, the operators now run on different nodes, in contrast to the previous plan. To specify, which operation is carried out at which node, every operation is tagged with the node's location specifier (prefixed by the @ symbol). Sometimes intermediate results have to be shipped from one node to another, for example, one input of the join operation. To reduce communication costs, the plan generator ensures that as many operations as possible are computed at the node where the data is stored. If an operation requires more than one input, like the join operator, the plan generator also makes sure that the smaller input is transferred over the network.

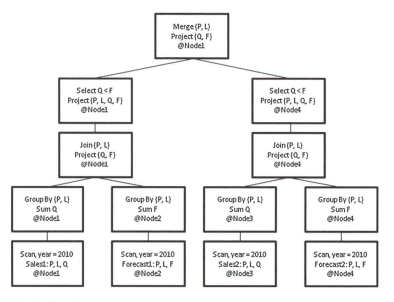

Fig. 4.15 Distributed query execution plan for the sales analysis query

Note that in our example, data partitioning (see Sect. 5.2) has a direct influence on the query performance. Because we partitioned the Sales and Forecast tables by P and L, the parallel joins, the aggregation, and the selections are safe (that is, no tuples are left out). If this were not the case, further intermediate results would need to be shipped from one node to another. In our example, the merge is just a concatenation operation, because the P, L pairs are disjoint by assumption. A real merge (with value-based comparisons on P and L) would be more expensive.

4.3 Compression for Speed and Memory Consumption

The largest real-world database installations have to handle up to several PB of data [123]. Despite the growing capacities of main memory in recent computer systems, it is still very expensive to enable in-memory processing for such data volumes. Therefore, storage engines of modern in-memory databases need to use compression techniques.

Column-oriented databases are particularly well suited for compression because data of the same type is stored in consecutive sections. This makes it possible to use compression algorithms specifically tailored to patterns that are typical for the data type [1].

An additional benefit coupled to compression is the potential for improved query performance. This applies especially to the case of light-weight compression where the CPU overhead for compressing/decompressing does not outweigh the gain of a

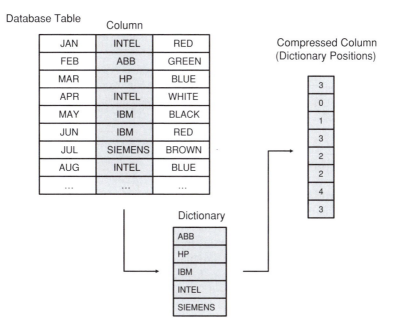

Fig. 4.16 Dictionary encoding

logically higher memory bandwidth [25, 184]. While the current hardware trend (see Sect. 1.3) induces a growing gap between memory bandwidth and CPU performance, there are indications that this trend is also causing a shift towards the usage of heavier-weight compression techniques within (in-memory) databases [1].

In the following section, we describe some light- and heavy-weight compression schemes that can be applied to the columns of a database. Finally, we give results of compression analysis on real customer data.

4.3.1 Light-Weight Compression

Most of the light-weight compression methods are based on dictionary coding, that is, expressing long values using shorter ones [1, 105, 184]. In practice, every distinct value of a column is added to a dictionary, and each occurrence of that value is replaced by the position of its dictionary entry [105]. For an example see Fig. 4.16. Using this representation of a column, different techniques that are described below can be applied to shorten the sequence of dictionary positions. In addition, compression can also be applied to that modified sequence [1, 149, 105] as well as to the dictionary itself [14].

While full-table scans are straightforward to implement on light-weight compressed columns, selecting single rows is often more difficult as most schemes do not allow direct random access to the compressed cells. Therefore, we will give details on how to reconstruct single entries when using the presented compression schemes.

Note also that implementation details, such as memory-alignment or the use of low-level instructions, are crucial for compression and decompression performance [148, 185]. The light-weight compression techniques introduced here are also used by SanssouciDB.

Common Value Suppression

In many real-world scenarios there are database columns that contain only a small fraction of non-null values [148]. In these cases, the column has to store many repetitions of the same symbol. Common value suppression techniques try to exploit such a redundancy. For example, prefix coding replaces consecutive identical values at the beginning of a column by storing the number of the repetitions [105]. Sparse coding achieves good compression if a predominant value is scattered among the entire column. It stores one bit for each cell indicating whether it contains an exception value that has to be stored explicitly. See Fig. 4.17 for an example. To retrieve the value for a given row index, it is firstly required to determine whether the row contains the sparse value or not. This can be done in constant time by just checking the bit at the given row position. In case the bit is not set, the row actually contains the sparse value. Otherwise, there is an exception value that has to be retrieved. Its position or rank in the sequence of exception values is given by the number of bits that are set up to the respective row index. In practice, rank and select data structures can be used to efficiently implement calculating this position. Using such a data structure, any single value of a sparse-coded column can be extracted in constant time.

For the specific case where the predominant value is null, these techniques are also embraced by the term null suppression [1].

Run-Length Encoding

This encoding scheme exploits runs of the same value within a column [1, 105]. Logically, it replaces the runs by the value of the runs and the number of repetitions within that run. In practice, we store the last row index of each run instead of the number of repetitions (see Fig. 4.18). This allows us to retrieve any cell value quickly by just determining the position of its run, for example, using binary search, and reading the correspondent value. Thus, the value for a given row number can be extracted in logarithmic time in the number of runs.

Run-Length Encoding (RLE) can be considered as a more generalized form of common value suppression, which is particularly good for sorted columns [1].

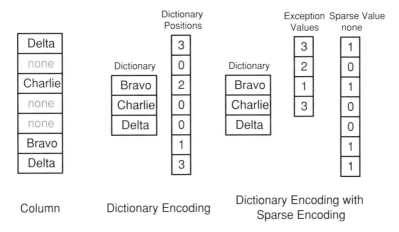

Fig. 4.17 Sparse encoding

Cluster Coding

Cluster coding works on equal-sized blocks containing only few distinct values [105]. Blocks with a single distinct value are compressed by storing just that value in a separate sequence. An additional indicator for each block stores whether the block is compressed or not. For a given row, the index of the corresponding block indicator can be obtained by the integer division of the row number by the block size. Selecting the correct block from the sequence of uncompressed and explicitly stored blocks works analogically to retrieving exception values for sparse coding. That is, counting all compression indicators containing a one up to the given block number yields the correct location of that uncompressed block. As mentioned earlier this count is also called the rank of the block. The sequence position of the single value of a compressed block is consequently given by the block index minus the block rank. Using rank and select data structures cited earlier, any single value of a cluster-coded column can be retrieved in constant time.

Indirect Coding

Similar to cluster coding above, indirect coding operates on blocks of data [105]. Additional dictionaries on block level further narrow down the range of values to be coded and the sizes of the dictionary entries. Actual column values can only be accessed indirectly because sequences of integers must first be resolved to the position of the global dictionary entry that points to the cell content. Generally, each block can have its own dictionary, but subsequent blocks can also share a dictionary, if they do not increase the memory footprint of the dictionary entries already stored. If a block has too many different entries it can also use the global dictionary directly

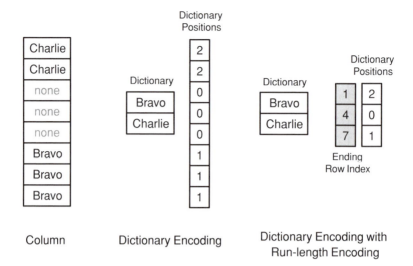

Fig. 4.18 Run-length encoding

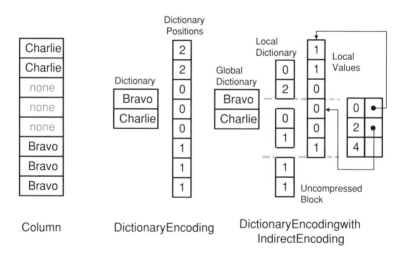

Fig. 4.19 Indirect coding

(Fig. 4.19). In practice, we replace each block to be compressed with a list containing all distinct values that appear in this block serving as the local dictionary. In addition, we store for each locally compressed block the sequence of local dictionary positions. For reconstructing a certain row, an additional data structure holds a pointer to the (local or global) dictionary and the corresponding position sequence for each block. Thus, the running time for accessing a row is still constant.

It is also possible to use combinations of the approaches shown above. For example, cluster coding can be combined with null suppression using prefix coding [105]. A column with many distinct values would use dictionary coding alone—without any modification to the sequence of dictionary positions. As mentioned above, compression can also be applied to the shortened sequences of dictionary positions. However, indirect coding is useless if the (modified) position values are stored uncompressed, each using the full width of a built-in integer data type. Some of the most popular techniques reducing value widths are described in the following.

Bit Compression

Instead of storing each value using a built-in integer data type, bit compression uses only the amount of bits actually needed for representing the values of a sequence. That is, the values are stored using fixed-width chunks, whose size is given implicitly by the largest value to be coded [150].

Variable Byte Coding

Compared to bit compression, variable byte coding uses bytes rather than bits as the basic unit for building memory chunks [118]. Generally, all values are separated into pieces of seven bits each, and each piece is stored in a byte. The remaining leading bit is used to indicate whether this is the last piece of a value. So, each value can span a different number of bytes. Because the memory of almost all modern computer systems is byte-addressable, compressing and decompressing variable byte codes is very efficient. It is also possible to generalize variable byte coding using chunks with sizes other than eight bits [179].

Patched Frame-of-Reference

Patched Frame-of-Reference (PFOR) compression groups the values to be coded in fixed-sized memory blocks [190]. All values within a block are stored as offsets from a previously defined block base. The offsets are bit compressed where the width parameter is not necessarily chosen such that it fits each value. Instead, so-called exceptions are allowed. These exception values are stored uncompressed in reverse order beginning at the end of each block.

4.3.2 Heavy-Weight Compression

One of the most popular heavy-weight compression schemes is Lempel-Ziv encoding. The basic idea is to replace repeatedly occurring patterns by references to their previous appearances. The non-overlapping patterns are stored in a (self-referencing) table that is built dynamically while parsing the input data. The entries in the table are distinct, and each new entry (except the first) is a reference to an existing one combined with an additional character (or byte). Lempel-Ziv coding can be advantageously used in database systems [184]. But in general, heavy-weight compression needs more compression and decompression effort than light-weight compression. As already discussed above, the use of heavy-weight compression schemes has to be carefully considered.

Another option is to use a hybrid approach of heavy- and light-weight schemes, compressing less frequently accessed data more than current content.

4.3.3 Data-Dependent Optimization

The compressibility induced by encoding schemes is highly dependent on the characteristics of the data. As well as carefully choosing the most suitable compression technique, it is possible to further restructure the data to better fit the selected scheme [83, 75]. For example, RLE can compress best if a column is first reordered such that there is only a single run for each distinct symbol. However, as soon as more than one column is involved, this is a non-trivial problem because corresponding entries among other columns have to be placed at equal positions in each column after reordering. For run length encoding, this problem is NP-hard [6]. Hence, there are heuristics to tackle the problem of reordering in combination with the selection of the best compression scheme [105].

4.3.4 Compression-Aware Query Execution

Compression can be applied to databases by abstracting the compression methods to higher-level operations [1]. Compression can also be exploited to reduce querying times. Even though most compression schemes do not allow direct random access to single cells of a column, specialized implementations, working directly on the compressed data of a certain method, can lead to significant performance improvements [1, 105, 106]. For example, a select query on the predominant value of a sparse coded column is—for all intents and purposes—free of cost: the sequence of bits indicating whether a row contains the sparse value or not is already the result.

In addition, index structures can be used on top of compressed data to reduce the overhead of unpacking too much unused data. For example, accessing a certain value in a variable-length encoded column requires unpacking all its predecessors because the position of that value is not known in advance. However, it is often possible to store synchronization points allowing the process to proceed from a point close to the requested value. Compression can also be applied to the indices themselves [180].

Table 4.2 Compression factor on selected columns of accounting table

	Sparse	RLE	Cluster	Indirect
Amount	2.5	2.7	2.9	4.9
Debit/Credit indicator	7.6	10.2	14.5	11.6
Date	222.3	1948.1	99.4	231.1
Discount	113	103.1	98.7	106.2

The advantages of dictionary coding are manifold. First, the dictionary maps more complex data types such as strings to simple integers. Thus, querying the data can largely be done on these integers allowing the use of the highly optimized Arithmetic Logic Unit (ALU) of the CPU. Searching for a single value within a column can be done by mapping the value to its dictionary position and scanning the position sequence linearly using simple integer comparisons. The entries of the dictionary can be stored sorted so that interval searches can be implemented by using just two integer comparisons while scanning the positions. The first comparison is required to check the lower interval bound, and the second is needed to check the upper interval bound.

4.3.5 Compression Analysis on Real Data

We have analyzed real company data to determine the compression that can be achieved using light-weight schemes. Table 4.2 shows compression factors for selected columns of the accounting table of a large ERP system. The numbers show that it is actually useful to exploit the variety of compression schemes according to the characteristics of the data.

For example, the amount column has the highest cardinality, and therefore, it is less compressible than all other listed columns. Nevertheless, indirect coding can reduce its size by approximately 80%. The debit/credit indicator column encodes the type of the transaction. Cluster coding can exploit uniform blocks containing a single transaction type, but it can also store interleaved runs without too much overhead. Because the number of rows in the accounting table increases chronologically, the date column contains many runs with recurring dates. Therefore, RLE can achieve enormous compression. The discount column has zero as the predominant value. Hence, sparse coding works best here.

The overall compression factor that could be achieved for the accounting table was 50. Given that systems with several TB of main memory per blade are already available at the time of writing, even large databases with up to several PB of data can be run completely in main memory.

4.4 Column, Row, Hybrid: Optimizing the Data Layout

To support analytical and transactional workloads, two different types of database systems are used. A distinct property of OLAP applications is that if tables are joined together based on their relations the join cardinality is much higher compared to OLTP operations. Thus OLAP systems provide approaches to solving this problem by specialized join algorithms or denormalization.

Based on the knowledge of the memory hierarchy one approach to avoid performance problems for combined analytical and transactional applications, is to vertically partition the tables used by the given application depending on the workload. With such an approach it would be possible to optimize the access to the physical data storage and to increase the overall performance of the system.

In this context where properties of traditional row- and column-oriented databases are mixed, we call such data storages hybrid databases to reflect the combination of the other approaches.

4.4.1 Vertical Partitioning

Vertical partitioning means splitting the column set of a database table and distributing it across two (or more) tables. The same operation is performed during normalization, but for performance or availability reasons, it can be advantageous to partition even a normalized table. Key to success is a thorough understanding of the application's data access patterns. Attributes that are usually accessed together should be in the same table, as every join across different tables degrades performance.

In the history of databases the storage was typically optimized for so-called point operations where a single tuple is selected and returned. One typical example is to display the details of a given order or customer identified by the primary key of the table. The primary key is then indexed and can directly point to the stored tuple. For in-memory database systems this can be even easier since the index may hold the direct offset to the tuple in memory. Figure 4.20 shows such an operation where a single tuple is retrieved using a row store (left part) and column store (right part). In addition, for the row and column store, the arrow indicates the direction of storage—increasing addresses and the width of the cache line, assuming that eight attributes fit into one cache line.

To compare the access performance of both storage models—for this single tuple operation—we can make a first assumption: when using the row store, only one cache miss occurs for the retrieval of the data. This results when the attribute a1 is read and due to the implementation of the memory hierarchy the attributes a2 to a8 that reside on the same cache line are read, too. For the column store the situation is different: since each column persists all tuples of a single attribute data is loaded into the CPU cache that is not accessed. For each accessed attribute one cache line containing all

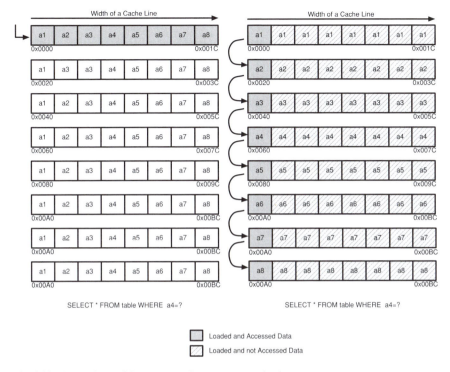

Fig. 4.20 Comparison of data access using row store and column store

tuples of this attribute is loaded whereas only one tuple is actually accessed. As a result eight cache misses occur.

Similar behavior can be observed for other database queries. In the second example an analytical query aggregates on one attribute and evaluates a predicate on a second column. In contrast to the first query the selectivity on both attributes is very high; it is not uncommon that all tuples from those two attributes—the aggregating column and the predicate column—are accessed. Figure 4.21 shows the access patterns for this operation.

Again the dark gray color on attributes a_1 and a_4 marks the tuples that are accessed and the arrow indicates the width of the cache line and the increasing addresses. For the row store this operation is a clear disadvantage: every row is accessed and a full cache line is loaded. Every attribute that resides on the cache line is loaded although for the row store attributes a_2, a_3, a_5, a_6, a_7, and a_8 are never read—meaning for every row in this simple case 75% of the loaded data is not accessed and wasted. The situation is different for the column store where only the two columns a1 and a4 are accessed—the other attributes still reside in memory but will not be accessed. Both of the above examples show that given a certain workload one of the physical storage models is more appropriate than the other one. Enterprise applications typically

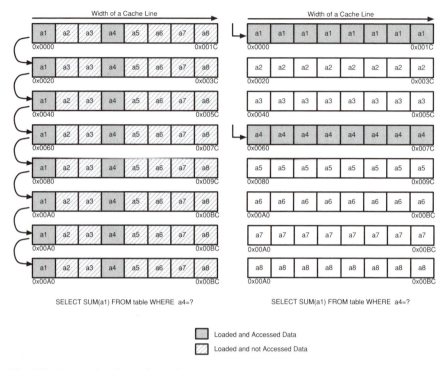

Fig. 4.21 Accessed attributes for attribute scan

belong to the transactional workload category and would issue mostly queries that are optimal for the row store.

An example of a vertical partitioned layout is shown in Fig. 4.22. The question is how can we evaluate if the layout is appropriate for the given workload? To perform the comparison, a common measure is required, and based on this measure; the best layout can be chosen. One possibility is to count the cache misses for each query and layout in the workload. As shown in the Sect. 4.1, each cache miss induces a stall where the CPU has to wait for the new values to be accessible.

Table 4.3 compares the number of cache misses for each given layout above with the transactional workload, the analytical workload and the mixed workload. In this example it is assumed that a single cache line holds eight values of a fixed length type and the width of the table is 8 attributes (one cache line) and the table has eight rows. The number of cache misses shows that the hybrid layout is a global optimization with regard to the mixed workload. For the transactional query it produces three cache-misses instead of one for row-oriented storage, but compared to column-oriented storage it produces almost a factor three less cache misses. A similar picture is drawn for the analytical query: here, the hybrid storage structure generates a factor four less cache misses compared to row-oriented storage and generates the same amount of cache misses as column-oriented storage.

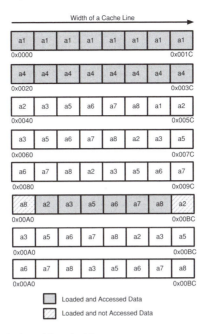

Fig. 4.22 Example of vertical partitioned table

Table 4.3 Comparing cache misses for different workloads and storage types

Transactional misses	Analytical misses	Mixed workload misses	Workload
Row store	1	8	9
Column store	8	2	10
Hybrid store	3	2	5

In this very simple example it is easy to determine the most appropriate layout, but the situation becomes more complicated as more queries participate in the workload and the complexity of the queries increases. In real applications the number of possible layouts explodes: for a table with five attributes there are already 73 different layouts and for a table of ten attributes 4,596,553 different layouts. The number of possible layouts corresponds to the number of partitions $\{1, \ldots, n\}$ into any number of ordered subsets. For a table with n attributes there are $a(n)$ different hybrid layouts (recursive formula).

$$a(n) = (2n - 1)a(n - 1) - (n - 1)(n - 2)a(n - 2) \qquad (4.4)$$

The solution space of possible layouts is huge and finding an optimally portioned layout is NP-hard. Any algorithm that guarantees to determine the optimal layout will run in exponential time.

Fig. 4.23 Relational operator
tree for a simple query

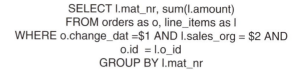

SELECT l.mat_nr, sum(l.amount)
FROM orders as o, line_items as l
WHERE o.change_dat =$1 AND l.sales_org = $2 AND
o.id = l.o_id
GROUP BY l.mat_nr

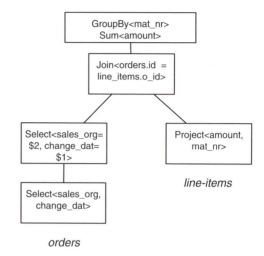

To summarize the above findings, it is possible to use cache misses as an optimization criterion, since cache misses can be translated into stall cycles that can be translated into time. When using vertical partitions, the large number of possible layouts does not allow for manual selection of the appropriate layout by the database administrator but instead requires automatic layout selections.

4.4.2 Finding the Best Layout

The goal of this section is to present a possible approach to determine the optimal layout of a given workload. To optimize the access cost in the context of hybrid databases, it is essential to know the workload in advance. As a second step all distinct queries are analyzed with regards to their cache miss behavior. Together with the weight of the query, the access cost is the foundation in which to evaluate whether or not a certain layout is better.

To estimate the costs of a query, it is not only necessary to know the query itself, but also to understand how it is evaluated by the database system.

Figure 4.23 shows a relational operator tree. In this example there are two tables: orders and line_items. One storing the sales order headers and the other stores the sales order line items. The operator tree is then evaluated bottom-up. Every operator is executed and produces the input for the next. Data flows from the bottom to the top where it is returned as the result of the query. The operator tree is a high level description of how the query is evaluated. The functionality of an operator is usually implemented in the native language of the DBMS. The description of the implementation in a form that can be used to estimate the execution costs is difficult. One distinct way to describe the implementation is using access patterns as shown in related work. One representative for a cost model based on access patterns is explained in [110]. Here, the authors describe the influence of those access operators in detail and how the different access patterns can be combined.

Estimating the costs of a query is necessary for query as well as layout optimization. It is generally desirable to estimate the costs in a measure that has a total order (for example, a simple integer value). This allows the comparison of two values in the metric and determining which one is better. To calculate this value, the model may use any number of intermediate measures. To find the best possible layout it is necessary to perform a cost analysis directly on the operator tree to take intermediate results and repetitive access into account. Since the relational operator tree is such a model, it can be used for a more accurate cost estimation. However, for cost calculation the expected input and output sizes have to be calculated. Those sizes depend on many different variables, such as value distribution and selectivity of the predicates.

We consider a query workload W, consisting of a set of queries $q_i : W = \{q_1 \ldots q_m\}$ that are regularly posed against the database. Each query has a weight that captures the relative frequency of the query. Furthermore, each query has a cost $Cost_{DB}(q)$, representing the time required by the database to answer the query. The time needed to answer all queries of a given workload W is proportional to the following expression:

$$Cost_{DB}(W) \, \Sigma_{i=1}^{m} w_i \cdot Cost_{DB}(q_i) \tag{4.5}$$

where $Cost_{DB}(q_i)$ is the cost associated with the operations performed as part of q_i combined with the weight of this query. Formally, given a database DB and a workload W, our goal is to determine the list of layouts λ_{opt} minimizing the workload cost:

$$\lambda_{opt} = argmin_\lambda(Cost_{DB}(W)) \tag{4.6}$$

When the output tuples can be reconstructed without any cache eviction, the cost expression distributes over the set of queries in the sense that each cost can be decomposed and computed separately for each partition and the corresponding subsets of the queries accessing the partition. Based on a detailed cost model and the above observations, our layout algorithm works in three phases called candidate generation, candidate merging, and layout construction phases. Similar approaches are described in [76] and [122]. In contrast to the aforementioned methods, our layout generation algorithm is optimized for main memory systems and not for optimized I/O in disk based database systems.

The first phase of our layout algorithm determines all primary partitions for all participating tables. A primary partition is defined as the largest partition that does not incur any container overhead cost. Container overhead cost is described as the cost when additional attributes are loaded into the cache, which are not accessed by subsequent operators. Each operation op_j implicitly splits this set of attributes into two subsets: the attributes that are accessed by the operation and those that are ignored. The order in which we consider the operations does not matter in this context. By recursively splitting each set of attributes into subsets for each operation op_j, we end up with a set of K primary partitions each containing a set of attributes that are always accessed together. The cost of accessing a primary partition is independent of the order in which the attributes are laid out, since all attributes are always queried together in a primary partition.

The second phase of the algorithm inspects permutations of primary partitions to generate additional candidate partitions that may ultimately reduce the overall cost for the workload. As shown using our cost model, merging two primary partitions P_i^1 and P_j^1 is advantageous for wide, random access to attributes since corresponding tuple fragments are co-located inside the same partition; for full projections, the merging process is usually detrimental due to the additional access overhead. This tension between reduced cost of random accesses and penalties for large scans of a few columns allows us to prune many of the potential candidate partitions early in the process. We compute the cost of the workload W, $Cost_{P_i^n}(W)$ on every candidate partition P_i^n obtained by merging n primary partitions $(P_1^1, \ldots, P_n^1)$, for n varying from 2 to $|P|$. If this cost is equal or greater than the sum of the individual costs of the partitions (due to the container overhead), then this candidate partition can be discarded: in that case, the candidate partition can systematically be replaced by an equivalent or more optimal set of partitions consisting of the n primary partitions $P_1^1, \ldots, P_n^1$. If a candidate partition is not discarded by this pruning step, it is added to the current set of partitions and will be used to generate valid layouts in the following phase. A layout is valid if it consists of all attributes of the original partition.

The third and last part of our algorithm generates the set of all valid layouts by exhaustively exploring all possible combinations of the partitions returned by the second phase. The algorithm evaluates the cost of each valid layout consisting of a covering but non-overlapping set of partitions, discarding all but the physical layout yielding the lowest cost. The worst-case space complexity of our layout generation algorithm is exponential with the number of candidate partitions. However, it performs very well in practice, since very wide relations typically consist of a small number of sets of attributes that are frequently accessed together (creating only a small number of primary partitions) and since operations across those partitions are often relatively infrequent (drastically limiting the number of new partitions generated by the second phase above). [67] shows how to apply automated vertical partitioning for mixed enterprise workloads. Depending on the query distribution of the workload the hybrid storage design can achieve up to 400% performance improvements compared to pure column or row-oriented storage layouts.

4.4.3 Challenges for Hybrid Databases

As shown previously such a hybrid main memory storage engine is better suited for mixed workloads than a pure row or column store. Some remaining challenges have to be addressed additionally: one of the major problems is the possible change of workload. The chosen layout algorithm has to be robust to avoid frequent layout changes on a physical level, since reorganization is a very expensive task and can be compared with index reorganization operations. Typically, enterprise workloads are more or less static and the change in queries can be associated with version changes of the application software. Again, it becomes more important to combine semantic information from the application with the persistence layer. Due to the fact that layout changes will not happen from one second to another and probably are bound to release changes, it is possible to calculate an optimal transformation from the old physical layout to the new one.

In addition to the workload, other topics may have further impact on the performance. To increase the bandwidth between CPU and main memory, data can be compressed. Compression will change the consumed space and requires different calculation of the accessed data regions (Sect. 4.3). In addition the combination of different compression algorithms and sort orders on different partitions adds more complexity to the layouter and query execution, but promises better execution

4.4.4 Application Scenarios

The advantage of a hybrid database lies in the flexibility of the storage layer. While traditional row- or column-oriented databases are optimized for a specific workload, hybrid databases can adapt to any workload.

Take, for example, the following simple scenario: during sales processing in an enterprise point of sales (POS) application, data is stored in a single table. This application requires high insert performance, so the table should be organized in a row-oriented manner. To meet the requirements of both access types, the workload needs to be evaluated and split into the separate access patterns of each query as shown in Fig. 4.23. In this simple case we can assume the following constraints: since we want to perform fast filtering and aggregation on the `date` and `value` columns we need to split those two attributes from the rest of the table. This was shown earlier in Fig. 4.22. Compared to a pure column-oriented database, the data structure is still optimized for high-throughput write operations but supports fast full attribute scans on the predicate attributes. Optimizing the system for the mix of both operations would not be possible otherwise.

4.5 The Impact of Virtualization

Virtualization is mainly employed to (a) increase the utilization of a lightly loaded system by consolidation or to (b) increase the ease of administration due to the possibilities of rapidly provisioning a Virtual Machine (VM) or migrating VMs. Both are crucial for efficiently managing large data centers. As discussed in the previous sections of this chapter, achieving good cache hit rates by reading sequentially in main memory is a key component in maximizing query performance. Hardware prefetching provided by modern CPUs helps to further maximize cache hit rates. It is, however, not obvious how the kind of resource sharing introduced by virtualization affects the performance of an in-memory database. In the following, we present an experiment that was first published in [69].

4.5.1 Virtualizing Analytical Workloads

Column-oriented databases are well known to be suitable for analytic workloads [164]. Since this book focuses on in-memory databases, we use a prototypical implementation of SanssouciDB [135, 155] for our experiments. We are interested in understanding the effects of virtualizing the CPU and the physical system's main memory in the case of a in-memory column store. For this purpose, we run a multi-user version of the TPC-H benchmark [177] from the Transaction Processing Performance Council (TPC) [36], which will be described in the next paragraph. Finally, we present our findings.

4.5.2 Data Model and Benchmarking Environment

For the experiments presented here we adapted the Star Schema Benchmark (SSB) [126], a modified version of the TPC-H benchmark, which has been adjusted for multi-user OLAP workloads. At the level of the data model, the most important differences between TPC-H and SSB are the following:

- The TPC-H tables lineitem and orders are combined into one table called lineorders. This change transforms the TPC-H data model from 3rd Normal Form (3NF) into a star schema, which is common practice for data warehousing applications.
- The TPC-H table partsupp is dropped because it contains data on the granularity of a periodic snapshot, while the lineorder table contains data on the finest possible granularity: the individual line items. It is sufficient to store the data on the most fine-grained level available and to obtain numbers on a coarser granularity by aggregation.

While TPC-H has 22 independent data warehousing queries, SSB has four query flights with three to four queries each (13 queries in total). A query flight models a drill-down—that is, all queries compute the same aggregate measure but use different filter criteria on the dimensions. This structure models the exploratory interactions of users with BI applications. We modified SSB so all queries within a flight are performed against the same transaction ID to ensure that a consistent snapshot is used.

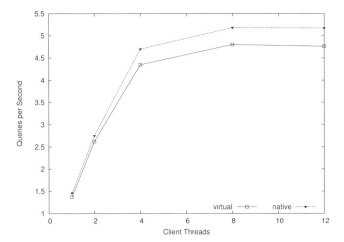

Fig. 4.24 Basic overhead of virtual execution

We also extended SSB with the notion of tenants and users. While multi-tenancy can be implemented on multiple levels (that is, shared machine, shared process, and shared table [91], see also Sect. 8.3), we chose the shared machine variant for the experiments presented here. With shared machine multi-tenancy, every tenant either gets his or her own database process or a distinct virtual machine. For this experiment, the latter was chosen. As we shall see in the remainder of this section, the limiting factor when scaling the number of concurrently active VMs is the bandwidth between CPU and main memory. We expect the results to also apply to shared process architectures. In our particular setup, each tenant has the same number of rows in the fact table.

4.5.3 Virtual Versus Native Execution

As a simple experiment, we ran the workload with one single tenant natively on the machine and increased the number of parallel users for our tenant from one to twelve. We then ran the same experiment inside a Xen virtual machine on the same server, which was configured such that it can use all the resources available on the physical system (CPU and main memory). Note that size of the tenant in compressed columns was significantly smaller than the 32 GB available on the machine, so that no paging occurred even with the additional main memory consumed by the Xen processes enabling the virtualization.

Figure 4.24 shows that the throughput in terms of queries per second is 7% lower on average in a Xen-virtualized environment. We believe that this overhead is largely due to the fact that SanssouciDB needs to allocate main memory to materialize intermediate results during query processing. While read access to main memory is cheap in Xen, the hypervisor must be called when allocating memory. A column

Table 4.4 Averaged results across multiple VM configurations [69]

Configuration name	# VMs	# Users/VM	AVG resp. time (ms)	Queries/second
1 VM normal	1	2	811	2.46
1 VM high	1	3	1,170	2.57
1 VM overload	1	4	1,560	2.56
2 VMs normal	2	2	917	4.36
2 VMs normal + 1 VM burn	2	2	909	4.40
2 VMs normal + 2 VMs burn	2	2	916	4.37
3 VMs normal	3	2	1,067	5.57
4 VMs normal	4	2	1,298	6.16
4 VMs overload	4	4	2,628	6.09

scan is not guaranteed to result in a scan of a contiguous region in main memory since Xen uses shadow page tables to give the guest Operating System (OS) the illusion of a contiguous address space, although the underlying hardware memory is sparse [11]. The hypothesis that the overhead can be largely attributed to the handling of intermediate results is supported by our observation that the physical configuration spends 5.6% of the CPU cycles in the kernel (switch to kernel mode triggered by the SanssouciDB process), while the virtual configuration spends 6.2% cycles plus additional 3.5% cycles in the Xen layer. The function names called in those Xen modules suggest that they are concerned with memory management tasks.

4.5.4 Response Time Degradation with Concurrent VMs

A widespread reason for the use of virtualization is to increase the utilization of otherwise lightly loaded servers by consolidation. This increase in utilization does not come without a cost. After quantifying the basic overhead incurred by virtualization, we looked into the background noise resulting from running multiple virtual machines concurrently on the same physical host. We configured the Xen host system such that the physical machine was split into four VMs. Each VM was configured with two virtual CPU cores that Xen dynamically schedules across the eight CPU cores of the physical machine. Since the OLAP workload under investigation is CPU-bound given that enough bandwidth is available so that the CPUs never stall, we believe that it makes sense to slice the physical machine into VMs by distributing the CPU cores across the VMs. We are interested in understanding how the performance changes from the perspective of an individual VM when increasing the number of guest systems on the physical host running the same computation-intensive workload.

Table 4.4 shows our averaged results across the VM configurations that we have examined. Since each VM is assigned two virtual cores we consider a VM to be exposed to a normal load when two simultaneously active clients are permanently issuing scan queries against the VM. The reported values for average response times and throughput are averages across all client threads of all VMs that are concurrently

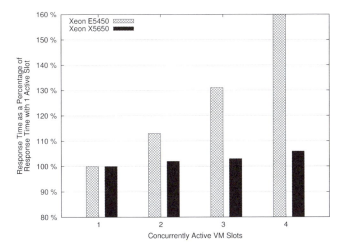

Fig. 4.25 Response time with increasing concurrency

running the benchmark at a particular point in time. When using only one out of the four available VM slots on the machine, the average response time for an OLAP query under normal load conditions is 811 ms. With a load that is slightly more than can be handled by the assigned CPU cores (that is, three client threads) the average response time is 1170 ms and degrades to 1560 ms when the number of client threads exceeds the number of processor cores by a factor of two (four client threads). The throughput in terms of queries per second is unaffected by the varying load. Note that we are interested in understanding the impact of concurrently active VMs when a physical system is partitioned into equally sized VMs. Thus, only two cores out of the eight ones available are used. A virtualized experiment using all available cores was shown in the previous section.

The impact of increasing concurrency in terms of simultaneously active VMs on average response times is also shown in Fig. 4.25. When running the SSB workload concurrently in two VMs the average response time is 917 ms across the four client threads (two per VM). In this configuration, two out of the four available VM slots are unused. An interesting variation of this configuration is to occupy the CPUs in the two spare VM slots. To do so, we use a tool consuming cycles on all available CPUs. However, this does neither affect the average response time nor the throughput measured across the two VM slots running the benchmark. Running the benchmark in the third and fourth slot additionally at normal load does however negatively impact the average response times (while the throughput increases as expected). Running all four slots in parallel at normal load results in an average response time of 1298 ms, which is approximately 60% higher when compared to using only one slot at normal load conditions.

We examined the 13 different queries in the benchmark in order to estimate how much data is transferred over the memory bus. We combined the average response time per query type with the sizes of the compressed columns of the dimension attributes on which the query has a predicate filter. We then factored in the size in MB of the requested fact table attributes multiplied with the selectivities achieved by the predicate filters. The selectivities for the SSB queries were taken from [126]. We then did a back-of-the-envelope calculation indicating that the benchmark requires up to 2 GB per second per active VM with two simultaneous users. All columns were fully compressed, that is, the differential buffer was empty (see also [135]). The Intel Xeon E5450 system under investigation has a memory bus bandwidth of 8 GB/s. We conclude that the above results suggest that the bandwidth of the memory bus is saturated and the CPUs stall on memory read when multiple VMs perform scans across contiguous memory regions at the same time.

The Intel Xeon E5450 processor, which was used in the previous experiments, is connected to a single off-chip memory controller via the system's FSB, which has been the prevalent architecture for many years. The Nehalem microarchitecture equips each processor with an integrated memory controller, and memory latency and bandwidth scalability are improved [116]. Consequently, we re-ran the experiment using an Intel Xeon X5650 system (Nehalem). Otherwise, the machine was configured in exactly the same way as the Xeon E5450 model.

As depicted in Fig. 4.25, the average response times increase only up to 6% when all four VM slots are concurrently running the benchmark with two simultaneous client threads per VM. Note that a similar behavior is likely to be observed when running four independent SanssouciDB processes on a single physical machine or even when running all tables for all tenants within the same SanssouciDB process; from the perspective of the memory bus the three alternatives should be similar. The results on the Nehalem platform indicate that when building a data center for multi-tenant analytics using an in-memory column database it might be more economic to use high-end hardware than commodity machines since more consolidation can be achieved without affecting end-user response times.

4.6 Summarizing the Technical Concepts

This section summarizes the technical concepts which have been taught so far.

Partitioning

We distinguish between two partitioning approaches: vertical and horizontal partitioning, whereas a combination of both approaches is also possible (Fig. 4.26). Vertical partitioning refers to the rearranging of individual database columns. It is achieved by splitting columns of a database table into two or more column sets. Each of the column sets can be distributed on individual databases servers. This can also be used to build up database columns with different ordering to achieve better search performance while guaranteeing high-availability of data. Key to the success of vertical partitioning is a thorough understanding of the application's data access patterns. Attributes that are accessed in the same query should reside in the same partition since locating and joining additional partitions may degrade overall performance. In contrast, horizontal partitioning addresses large database tables and how to divide them into smaller pieces of data. As a result, each piece of the database table contains a subset of the complete data within the table. Splitting data into equivalent long horizontal partitions is used to support search operations and better scalability. For example, a scan of the request history results in a full table scan. Without any partitioning, a single thread needs to access all individual history entries and checks the selection predicate. When using a naive round robin horizontal partitioning across 10 partitions, the total table scan can be performed in parallel by 10 simultaneously processing threads reducing response time to approx. one ninth compared to the single threaded full table scan.

Fig. 4.26 Partitioning

Minimal Projections

Typically, transactional enterprise applications follow a very simple access pattern (Fig. 4.27). A lookup for a given predicate is followed by reading all tuples that satisfy the condition. Interestingly, for traditional disk-based row databases it is very easy and fast to read all attributes of the table because they are physically co-located. Since the overall processing time is quite high due to the I/O overhead, it does not matter how many attributes are projected. However, the situation changes for in-memory column store databases. Here, for each selected tuple, access to each of the projected attributes will touch a different memory location, incurring a small penalty. Thus, to increase the overall performance, only the minimal set of attributes that should be projected for each query are selected. This has two important advantages: first, it dramatically reduces the amount of data that is transferred between client and server. Second, it reduces the number of accesses to random memory locations and thus increases the overall performance.

Fig. 4.27 Minimal projections

Reduction of Layers

In application development, layers refer to levels of abstractions. Each application layer encapsulates specific logic and offers certain functionality (Fig. 4.28). Although abstraction helps to reduce complexity, it also introduces obstacles. The latter result from various aspects, e.g. a) functionality is hidden within a layer and b) each layer offers a variety of functionality while only a small subset is in-use. From the data's perspective, materialized layers are problematic since data is marshaled and unmarshaled for transformation in the layer-specific format. As a result, the identical data is kept in various layers redundantly. To avoid redundant data, logical layers, describing the transformations, are executed during runtime, thus increasing efficient use of hardware resources by removing all materialized data maintenance task. Moving formalizable application logic to the data it operates on results in a smaller application stack and increases maintainability by code reduction. Furthermore, removing redundant data storage increases the audibility of data access.

Fig. 4.28 Reduction of layers

Reduction of layers

No Disk

For a long time the available amount of main memory on large server systems was not enough to hold the complete transactional data set of large enterprise applications. Today, the situation has changed. Modern servers can provide multiple terabytes of main memory and allow the complete transactional data to be kept in main memory. This eliminates multiple I/O layers and simplifies database design, allowing for high throughput of transactional and analytical queries (Fig. 4.29).

Fig. 4.29 No disk

No disk

Multi-Core and Parallelization

In contrast to the hardware development until the early 2000s, today's computing power no longer scales in terms of processing speed, but in the degree of parallelism.

Today, modern system architectures provide server boards with up to 8 separate CPUs where each CPU has up to 12 separate cores. For modern enterprise applications it becomes imperative to reduce the amount of sequential work and develop the application in a way that can be easily parallelized (Fig. 4.30).

Parallelization can be achieved at a number of levels in the application stack of enterprise systems: from within the application running on an application server to query execution in the database system. Processing multiple queries can be handled by multi-threaded applications meaning that the application does not stall when dealing with more than one query. Threads are a software abstraction that need to be mapped to physically available hardware resources. A CPU core can be considered as single worker on a construction area. If it is possible to map each query to a single core, the system's response time is optimal. Query processing also involves data processing meaning the database needs to be queried in parallel too. If the database is able to distribute the workload across multiple cores of a single system this is optimal. If the workload exceeds the physical capacity of a single system, multiple servers or blades need to be involved for work distribution to achieve optimal processing behavior. From the database perspective, the partitioning of data sets enables parallelization since multiple cores across servers can be used for data processing.

Fig. 4.30 Multi-core and parallelization

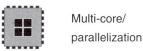

Multi-core/
parallelization

On-the-Fly Extensibility

The possibility of adding new columns to existing database tables on-the-fly dramatically simplifies a wide range of customization projects that customers of enterprise software are often required to do (Fig. 4.31). When physically storing consecutive tuples in row-oriented database, all pages belonging to a database table must be reorganized when adding a new column to the table. In a column store database, such as SanssouciDB, all columns are stored in physically separate locations. This allows for the simple implementation of column extensibility, without the need to update any other existing columns in the table. This reduces a schema change to a pure metadata operation, allowing for flexible and real-time schema extensions.

Fig. 4.31 On-the-fly extensibility

On-the-fly
extensibility

No Aggregate Tables

A very important part of the SanssouciDB philosophy is that all data should be stored at the highest possible level of granularity (e.g. the level of greatest detail). This is in contrast to the prevailing philosophy in most enterprise data centers, which says that the data should be stored on whatever level of granularity is required by the application to ensure maximum performance. Unfortunately, multiple applications use the same information and require different levels of detail, which results in high redundancy and software complexity around managing the consistency between multiple aggregate tables and source data. Given the incredible aggregation speed provided by SanssouciDB, all aggregates required by any application can now be computed from the source data on-the-fly, providing the same or better performance as before and dramatically decreasing code complexity which makes system maintenance a lot easier (Fig. 4.32).

Fig. 4.32 No aggregate tables

Analytics on Historical Data

All enterprise data has a lifespan: depending on the application, a datum might be expected to be changed or updated in the future often, seldom, or never. In financial accounting, for example, all data that is not from the current year plus all open items from the previous year can be considered historic data, since they may no longer be changed. In SanssouciDB, historical data is instantly available for analytical processing from solid state disk (SSD) drives. Only active data is required to reside in-memory permanently (Fig. 4.33).

Fig. 4.33 Analytics on historical data

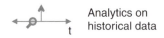

Combined Row and Column Store

To support analytical and transactional workloads, two different types of database systems evolved (Fig. 4.34). On the one hand, database systems for transactional workloads store and process data in rows, i.e. attributes are stored side-by-side. On the other hand, analytical database systems aim to analyze selected attributes of huge data sets in a very short time. If the complete data of a single row needs to be accessed, storing data in a row format is advantageous. For example, when comparing details of two customers, all database attributes of those customers, such as name, time, and content need to be loaded. In contrast, columnar databases benefit from their storage format when a subset of attributes needs to be processed for all or a huge number of database entries. For example, summing up the total number of products that passed a certain reader gate involves the attributes date and business location while ignoring the product id and the business step. Using a row store for this purpose would result in processing all attributes of the event list, although only two attributes are required. Therefore, incorporating a columnar store benefits from accessing only relevant data and fewer search skipping operations.

Fig. 4.34 Combined row and column store

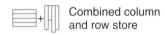

Combined column and row store

Object Data Guides

The in-memory database improves the performance of retrieving a business object by adding some redundancy to the physical data model. This redundancy represents a join index for querying sparse tree-shaped data, which is called the object data guide and includes two aspects: in addition to the parent instance, every node instance can contain a link to the corresponding root instance. Using this additional attribute, it is possible to retrieve all nodes in parallel instead of waiting for the information from the parent level. Additionally, each node type in a business object can be numbered. Then, for every root instance, a bit vector, the Object Data Guide, is stored, whose bit at position i indicates whether an instance of node number i exists for this root instance. Using this bit vector, a table only needs to be checked if the corresponding bit is set, reducing the complexity of queries to a minimum. Furthermore, the amount of data retrieved and transmitted is minimized as well (Fig. 4.35).

Fig. 4.35 Object data guides

Object to relational mapping

Group-Key

A common access pattern for enterprise applications is to select a small group of records from larger relations, for example, all line-items belonging to an order (Fig. 4.36). The standard execution of such an operation scans the complete table and evaluates the selection condition for every record in the table. Applications executing such operations frequently suffer from a degraded performance since the complete table is scanned often, although only a small group of records match the selection condition. To speed up such queries, group-key indexes can be defined built on the compressed dictionary. A group-key index maps a dictionary-encoded value of a column to a list of positions where this value can be found in a relation.

Fig. 4.36 Group-key

Group-Key

MapReduce

MapReduce is a programming model to parallelize the processing of large amounts of data (Fig. 4.37). MapReduce has taken the data analysis world by storm, because it dramatically reduces the development overhead of parallelizing such tasks. With MapReduce, the developer only needs to implement a map and a reduce function, while the execution engine transparently parallelizes the processing of these functions among available resources. SanssouciDB emulates the MapReduce programming model and allows the developer to define map functions as user-defined procedures. Support for the MapReduce programming model enables developers to implement specific analysis algorithms on SanssouciDB faster. The parallelization and efficient execution is handled by SanssouciDB's calculation engine.

Fig. 4.37 MapReduce

MapReduce

Text Retrieval and Exploration

Elements of search in unstructured data, such as linguistic or fuzzy search find their way into the domain of structured data, changing system interaction, for example, enabling the specification of analytical queries in natural language (Fig. 4.38).

Furthermore, for business environments added value lies in combining search in unstructured data with analytics of structured data. Cancer databases in healthcare are an example where the combination of structured and unstructured data creates new value by being able to map (structured) patient data in the hospital database onto (unstructured) reports from screening, surgery, and pathology on a common characteristic, for example, tumor size and type, to learn from treatments in similar cases.

Fig. 4.38 Text retrieval and exploration

Bulk Load

Besides transactional inserts, SanssouciDB also supports a bulk load mode. This mode is designed to insert large sets of data without the transactional overhead and thus enables significant speed-ups when setting up systems or restoring previously collected data. Furthermore, bulk loading has been applied to scenarios with extremely high insert loads, such as RFID event handling and smart grid meter reading collection, by buffering events and then bulk-inserting them in chunks. While increasing the overall insertion rate, this buffered insertion comes at the cost of a small delay of data availability for analytics depending on the defined buffering period. This, however, is often acceptable for business scenarios (Fig. 4.39).

Fig. 4.39 Bulk load

SQL Interface on Columns and Rows

Business operations have very diverse access patterns. They include read-mostly queries of analytical applications and write-intensive transactions of daily business (Fig. 4.40). Further, all variants of data selects are present, including point selects, for example, details of a specific product and range selects, retrieving sets of data, for example, from a specified period like sales overview per region of a specific product for the last month. Column and row-oriented storage in SanssouciDB provides the foundation to store data according to its frequent usage patterns in a column or row-oriented manner to achieve optimal performance. Through the use of SQL, that supports column as well as row-oriented storage, the applications on top stay oblivious to the choice of storage layout.

Fig. 4.40 SQL interface on columns and rows

Any Attribute as an Index

Traditional row-oriented databases store tables as collections of tuples (Fig. 4.41). To improve access to values within specific columns and to avoid scanning the entire table indexes are typically created for these columns. In contrast to traditional row-oriented tables, the columnar storage of tuples in SanssouciDB allows the scanning of any columns corresponding to the attributes of the selection criteria to determine the matching tuples. The offsets of the matching values are used as an index to retrieve the values of the remaining attributes, avoiding the need to read data that is not required for the result set. Consequently, complex objects can be filtered and retrieved via any of their attributes.

Fig. 4.41 Any attribute as an index

Dynamic Multithreading within Nodes

Parallel execution is key to achieve sub second response time for queries processing large sets of data. The independence of tuples within columns enables easy partitioning and therefore supports parallel processing. We leverage this fact by partitioning database tasks on large data sets into as many jobs as threads are available on a given node. This way, the maximal utilization of any supported hardware can be achieved (Fig. 4.42).

Fig. 4.42 Dynamic multithreading within nodes

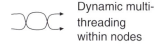

Dynamic multi-
threading
within nodes

Active and Passive Data Store

By default in SanssouciDB, all data is stored in-memory to achieve high-speed data access. However, not all data is accessed or updated frequently and so not everything needs to reside in-memory, as this increases the required amount of main memory unnecessarily. This so-called historic or passive data can be stored in a specific passive data storage based on less expensive storage media, such as SSDs or hard disks, still providing sufficient performance for possible accesses at lower cost. The dynamic transition from active to passive data is supported by the database, based on custom rules defined as per customer needs (Fig. 4.43).

We define two categories of data stores: active and passive: we refer to active data when it is accessed frequently and updates are expected. In contrast, we refer to passive data when this data either is not used frequently and neither updated nor read. Passive data is purely used for analytical and statistical purposes or in exceptional situations where specific investigations require this data. For example, tracking events of a certain pharmaceutical product that was sold five years ago can be considered as passive data for the following reasons: firstly, from the business perspective, the pharmaceutical is equipped with a best-before date of two years after its manufacturing date, i.e. even when the product is handled now, it can no longer be sold. Secondly, the product was sold to a customer four years ago, i.e. it left the supply chain and is typically already used within this timespan. Therefore, the probability that details about this certain pharmaceutical are queried is very low. Nonetheless, the tracking history needs to be conserved by law, for example, to prove the used path within the supply chain or when sales numbers are analyzed for building a new long-term forecast based on historical data.

Furthermore, introducing the concept of passive data comes with the advantage of reducing the amount of data which needs to be accessed in real-time, and to enable archiving. As a result, when data is moved to a passive data store it no longer consumes fast main memory and thereby frees hardware resources. Dealing with passive data stores involves the need for a memory hierarchy from fast, but expensive to slow and cheap. A possible storage hierarchy is given by memory registers, cache memory, main memory, flash storages, solid state disks, SAS hard disk drives, SATA hard disk drives, tapes, etc. As a result, rules for migrating data from one store to another need to be defined, we refer to this as an aging strategy or aging rules. The process of aging data, i.e. migrating it from a faster store to a slower one, is considered background task, which occurs on a regular basis, e.g. weekly or daily. Since this process involves reorganization of the entire data set, it should be processed during times with low data access, e.g. during nights or weekends.

Fig. 4.43 Active and passive
data store

Active/passive
data store

Single and Multi-Tenancy

To achieve the highest level of operational efficiency, the data of multiple customers can be consolidated onto a single SanssouciDB server. Such consolidation is key when SanssouciDB is provisioned in an on-demand setting (Fig. 4.44). Multi-tenancy allows makes SanssouciDB available to smaller customers at lower cost. Already today SanssouciDB is equipped with the technology to enable such consolidation while ensuring that no critical resources are contending between the customers sharing a server, and also ensuring a reliable and highly-available storage of the customer's data at the hosting site.

Fig. 4.44 Single and multi-
tenancy

Single and
multi-tenancy

Lightweight Compression

Compression is the process of reducing the amount of storage needed to represent a certain set of information. Typically, a compression algorithm tries to exploit redundancy in the available information to increase the efficiency of memory consumption. Compression algorithms differ in the amount of time that is required to compress and decompress data and the achieved compression rate defined as the reduction in memory usage. Complex compression algorithms will typically sort and perform complex analysis of the input data to achieve the highest possible compression rate at the cost of increased run-time (Fig. 4.45).

For in-memory databases, compression is applied to reduce the amount of data that is transferred between main memory and CPU, as well as to reduce overall main memory consumption. However, the more complex the compression algorithm is, the more CPU cycles it will take to decompress the data to perform query execution. As a result, in-memory databases choose a trade-off between compression ratio and performance using so called light-weight compression algorithms.

An example for a light-weight compression algorithm is dictionary compression. With dictionary compression, all value occurrences are replaced by a fixed length encoded value. This algorithm has two major advantages for in-memory databases: first, it reduces the amount of required storage and second, it allows the performing of predicate evaluation directly on the compressed data, thereby reducing the amount of data that needs to be transferred from memory to the CPU. As a result, query execution becomes even faster with in-memory databases.

Fig. 4.45 Lightweight compression

Insert-Only

Insert-only or append-only describes how data is managed when inserting new data

Insert-Only

Insert-only or append-only describes how data is managed when inserting new data (Fig. 4.46). The principle idea of insert-only is that changes to existing data are handled by appending new tuples to the data storage. In other words, the database does not allow applications to perform updates or deletions on physically stored tuples of data. This design approach allows the introduction of a specific write-optimized data store for fast writing and a read-optimized data store for fast reading. Traditional database systems support four operations for data manipulation; inserting new data, selecting data, deleting data, and updating data. The latter two are considered as destructive since the original data is no longer available after their execution. In other words, it is neither possible to detect nor to reconstruct all values for a certain attribute; only the latest value is available. Insert-only enables the storing of the complete history of value changes and the latest value for a certain attribute. This is also one of the foundations of all bookkeeping systems to guarantee transparency. For the history-based access control, insert-only builds the basis to store the entire history of queries for access decisions. In addition, insert-only enables tracing of access decisions, which can be used to perform incident analysis.

Fig. 4.46 Insert-only

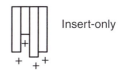

Insert-only

4.7 Conclusion

"Tape is dead, disk is tape, flash is disk, RAM locality is king", as Jim Gray put it in 2006 [63], probably best describes the current situation. With the shift in the memory hierarchy, it is now possible to keep the enterprise data of large companies in main memory. However, now the I/O bottleneck has shifted from the external-memory access gap to the main-memory access gap. In Sect. 4.1 we have shown the effects of caching with respect to the time to access a particular piece of data. Correctly allocating memory and placing data in the memory hierarchy—taking into account spatial and temporal locality—is of major importance for SanssouciDB. With even more innovations on the horizon, for example, non-volatile main memory, this topic will remain an exciting one.

To cope with the growth of enterprise databases, advances in memory technology as well as in processing power are needed. Due to the well-known physical limitations related to the clock speed of a single core CPU, multi-core processors and multi-blade systems have been introduced. To efficiently exploit this new processing power, the principles of parallel data processing were examined in Sect. 4.2 and how we exploit parallelism in SanssouciDB. To be able to scale with the existing problem size and hardware, and to make efficient use of existing processing resources, parallelization has to take place at all levels of abstraction: from parallelizing low-level SIMD instructions, intra-operator and inter-operator parallelism to large-scale parallelism in shared-nothing systems and to data centers.

Besides hardware developments, advances in software technology have made in-memory database systems possible. Compared to external-memory storage, main memory is still more expensive; therefore, compression techniques are required to gain a good trade-off between system cost and performance. We took a close look at various important compression techniques in Sect. 4.3. Of these, light-weight compression schemes, for example, common value suppression or cluster encoding are especially promising, because during query processing, the overhead for compressing/decompressing does not outweigh the gain of a logically higher memory bandwidth.

Another substantial development in database systems discussed in Sect. 4.4 is the capability of systems to store tables along rows, columns, or even both as combined columns. Combining these alternatives helps us to optimize the table data layout according to data access patterns in the query workload. Finally, in Sect. 4.5, we analyzed the impact of virtualization (that is, running our IMDB in a virtual-machine environment) on query processing. We have seen that the overhead of virtual execution versus native execution is less than 10%, in case only one instance of our DBMS in one virtual machine is running. When multiple DBMS instances were running on the same machine, each with its own virtual machine, we have shown the positive effects of the new Intel Nehalem architecture versus the predecessor model. Due to Nehalem's improved bandwidth and latency characteristics, response-time degradation was shown to be significantly lower. This supports our motivation to favor high-end hardware.

Achievements in hardware and software development have always had a major impact on database technology. Current advances in hardware lay the foundation for a new database approach that is still regarded by many people as radical: large-scale in-memory database systems for enterprise data. To develop such a system, detailed knowledge of the current hardware is essential.

Chapter 5
Organizing and Accessing Data in SanssouciDB

Abstract Providing enterprise users with the information they require, when they require it is not just a question of using the latest technology to store information in an efficient manner. Enterprise application developers and users need ways of accessing and storing their information that are suited to the tasks they wish to carry out. This includes things like making sure the most relevant data is always stored close to the CPU for fast access, while data that is no longer required for the day-to-day running of the business is stored in slower, cheaper storage; allowing new columns to be added to tables if customers need to customize an application; and allowing developers to choose the most efficient storage strategy for their particular task. The work of an application developer can also be made easier if they are able to read and write data in a way that fits in with the business process they are modeling. Finally, users and developers also need to have confidence that their database will be available when they need it, and that they will not lose their data if the power goes out or a hardware component fails. In this chapter we describe how data is accessed and organized in SanssouciDB to meet the requirements we have outlined. Enterprise applications place specific demands on a DBMS beyond those of just processing large amounts of data quickly. Chief amongst them is the ability to process both small, write-intensive transactions, and complex, long-running queries and transactions in a single system. As we have seen, column storage is well suited to the latter but it does not perform well on the single-row inserts which characterize write-intensive transactions. For this reason inserts can be directed to a write-optimized differential store and merged, when appropriate, with main storage. Scheduling these different workloads in an efficient manner is integral to the performance of the system. We do not want a long running query to block the entire system, but we also need to make sure that large numbers of small transactions do not swamp it. Another requirement that enterprise applications impose on a DBMS are that historical data must be kept for reporting and legal reasons, and it should be possible to store and access this data without severely impacting the performance of the system. The ability of enterprise applications to recover from failures without a loss of data and to have as little downtime as possible is also very important. In addition certain database

H. Plattner and A. Zeier, *In-Memory Data Management*,
DOI: 10.1007/978-3-642-29575-1_5, © Springer-Verlag Berlin Heidelberg 2012

operations, namely aggregation and join, form the majority of operations carried out by the database in enterprise applications, and special algorithms are required to take full advantage of the parallelization opportunities presented by modern multi-core processors.

5.1 SQL for Accessing In-Memory Data

SanssouciDB separates applications from physical details of data organization and memory manipulation. Instead of directly operating on data, the applications instruct SanssouciDB to perform certain tasks on their behalf. They do so by means of issuing SQL queries. SQL covers all aspects of data management in SanssouciDB: schema definition, manipulation and transaction control. SanssouciDB supports two alternative ways of working with SQL: sending queries to the query engine and calling stored procedures.

The goal of this section is to introduce SQL, the standard SanssouciDB API, and discuss different aspects of supporting it. In particular, the section starts with a general introduction of SQL and its role. Then, the lifecycle of an SQL query is described. After that, stored procedures are discussed. Finally, the section addresses the topic of data organization as it dramatically influences the performance of SQL queries.

5.1.1 The Role of SQL

One of the main features of SanssouciDB, and DBMSs in general, is that it isolates applications from the physical database. Rather than designing algorithms that iterate through tuples and manipulate them, application developers make statements in SQL. SQL is a high-level declarative language based on relational algebra and is designed for managing data in relational database management systems [48]. The language is divided into two parts: data definition and data manipulation. By using data definition constructs, developers can create and alter all types of database schema objects, for example tables, indices and views. By using data manipulation constructs, developers declare the actions to be performed on the database objects, for example insert, select, update, delete.

Listing 5.1 shows an example of a data manipulation SQL query that instructs SanssouciDB to select from the Customer table all names and identifiers corresponding to customers located in Europe, the Middle East and Africa.

Listing 5.1 (SQL Query)

```
SELECT name, id FROM Customer WHERE region = 'EMEA'
```

Listing 5.1 Example calling business function inside SQL

```
/** BFL Call **/
CALL BFLGROW(
    VALUES_TAB,
    GROWPERIODS_TAB,
    GROWRATE_TAB,
    GROWTYPE_TAB,
    SWITCHOVER_TAB,
    SWITCHOVERVALUE_TAB,
    RESULTS_TAB) WITH OVERVIEW ;
```

With declarative SQL queries an application developer only specifies the operations and data, on which the operations must be performed, and lets SanssouciDB determine the optimal way of their actual execution. In the example in Listing 5.1 the developer just instructs SanssouciDB to select only EMEA customers, without elaborating how the filtering must be executed. Depending on the selectivity of the predicate and the size of the table, SanssouciDB then decides on whether to perform a full table scan or use an index on the region column, if such an index exists. This contrasts to procedural programming style, where a developer is responsible for specifying all aspects of data processing. The flexibility and ease of declarative querying comes at the price of the overhead of query processing (parsing and optimization) [79].

5.1.2 The Lifecycle of a Query

In order to execute a query, SanssouciDB must perform a sequence of actions. The sequence starts with query parsing, which discovers what operations on what data must be performed. Having parsed the query, SanssouciDB constructs a query plan. Once the query plan is obtained SanssouciDB must resolve the logical schema into the primitives of the storage engine and execute the plan.

5.1.3 Stored Procedures

Stored procedures allow application developers to push down data-intensive application logic into the database layer. SanssouciDB stored procedures are programs written in procedurally extended SQL that are compiled and stored inside the database (Sect. 6.1). Stored procedures offer a number of advantages. First, they centralize business logic and facilitate its reuse. If many applications need to perform the same set of operations, a single stored procedure can be created and called by

Fig. 5.1 Example of a B+ tree

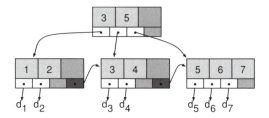

all applications. This reduces the amount of application code and simplifies change management. Second, stored procedures help to avoid network traffic and associated latency. Complex, repetitive tasks are often organized as sequences of SQL queries, where the result of one query gets processed in some way and becomes an input to another query. If this is done inside SanssouciDB, there is no need to transfer intermediate results and inputs between the database and an application. This reduces network traffic and network waiting time.

5.1.4 Data Organization and Indices

Tables are stored physically as collections of tuples. Depending on the use-case it is possible to optimize the organization of the collection to increase the performance of different access operations to the stored tuples [57]. The most common types of organizations are heaps, ordered collections, hashed collections and tree indices. Heap-organized collections place tuples in no particular order by appending a new tuple at the end of the collection. Inserting a new tuple is very efficient because new tuples are simply appended. Searching for a tuple in a heap-organized collection using any search condition requires sequential probing of all tuples of the collection. Because of its linear complexity, the search operation in heap-organized collections is considered to be expensive. The second possible data organization is an ordered collection. Ordered collections provide two advantages over heap-organized collections. First, reading the tuples in order is extremely efficient because no sorting is required. This is especially good in the context of sort-merge joins. Second, the search operation can be implemented more efficiently, for example, with a binary search that has logarithmic complexity. The advantages come at the price of the overhead of maintaining the sorted order of tuples in a collection. The third type of data organization is based on hashing. Hashing supports very fast access to the collection's elements on certain search conditions. The search condition must be an equality condition. The biggest advantage of hash collections is the ability to locate elements in constant time. This makes hashing particularly suitable for random access patterns. The biggest problem with hashing is the difficulty of discovering good hash functions, that is, functions that avoid collisions [95]. Hash-organized collections are often called hash indices.

The fourth data organization type is based on different tree types and is primarily used for secondary indices. This organization stores tuples using tree-like data structures, for example, a B+ tree. The tuples are arranged in a hierarchical data structure that is maintained by the DBMS. As in the case of sorted collections, tree-organized collections provide very good performance for search and range search operations. Figure 5.1 shows an example of a B+ tree. The leaf nodes of the tree contain the keys and pointers to the values while the non-leaf nodes contain the keys. The hierarchical structure allows searching for a key in logarithmic time while it is possible to iterate over the collection in sort order.

SanssouciDB stores all data in heap-organized collections. In case a particular query that is frequently executed is too slow and decreases the performance for the whole workload, SanssouciDB allows the creation of a group-key index on this attribute to speed-up predicate evaluation. The column store in SanssouciDB provides a special group-key index that directly builds on the compressed dictionary of the attribute. In contrast to traditional indices the group-key index stores the encoded value as a key and the positions where the value can be found as a value list. The group-key concept allows increasing the search performance for transactional and analytical queries significantly.

5.1.5 Any Attributes as Index

In contrast to traditional row-oriented databases, the column-orientation in SanssouciDB, by storing each column separately, facilitates scanning of single columns to filter values without having to read data that is not of interest. Thereby, the scan behaves in a similar way to a traditional index without having to create one explicitly since the sequential memory scan performance is fast enough to even apply a predicate on a non-indexed column. Consequently, complex objects can be filtered and retrieved via any of their attributes by using a full table scan as this is one of the basic operations in main memory column-store database systems.

As described in Sect. 4.3, columns of enterprise data generally consist of a low number of distinct values on a single column. Therefore, lightweight compression techniques, such as dictionary encoding, are applied to SanssouciDB in order to the reduce the amount of data to be processed and stored in main memory and all the more to leverage late materialization strategies during query execution. Since the dictionary compression replaces the actual values with an id that points to the dictionary of the column, the range of value ids has to be as big as that needed to represent the cardinality of the dictionary. In a basic dictionary compression a lot of compression opportunities are wasted by reserving 32 bits to define the value id space. Dictionary compression with bit-compressed value ids varies the length of the value ids and reserves only the needed number of bits to encode all distinct values but still guarantees fixed-length value ids. Given 15 values in the dictionary, the attribute vector that needs to be compressed needs only four bits to store each value id, allowing 128 value ids to fit on a 64 byte cache line. Similar to an order-preserving

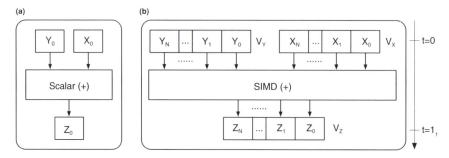

Fig. 5.2 SIMD execution model: in **a** scalar mode: one operation produces one result. In **b** SIMD mode: one operation produces multiple results

dictionary the disadvantage of this approach is that the bit-length of the value ids needs to be increased and all values in the attribute vector need to be rewritten when the number of distinct values in the dictionary exceeds the number of values that can be encoded with the current number of bits (see Sect. 5.5).

Since today's computer systems only address their memory in a byte oriented fashion there is no support in hardware to store values in a bit-compressed fashion. As a solution the information is encoded inside a larger data type, usually the size of a processor word, i.e. 64 bits on current machines. That means when a specific amount of bits needs to be addressed an offset is computed to represent that bit combination. These bit-wise operations are supported by current hardware with bit-oriented masking and shifting in every direction. As a result, in order to iterate on a bit-compressed vector the CPU has to shift for every row of the column to extract the exact bit-compressed value id. Since this is very compute intensive the database system will end up with a CPU-bound algorithm if no further optimization is applied.

Besides data parallelism on separate cores of a CPU, all modern CPUs support simultaneous execution operations on the elements of a vector. This concept is called Single Instruction stream–Multiple Data stream (SIMD). This concept was first classified by Flynn in [52] and represents a vector processing model providing instruction level parallelism. SIMD forms an important extension to modern processor architectures and provides the ability for a single instruction stream to process multiple data streams simultaneously. Figure 5.2 shows the SIMD execution model.

To process a number of compressed values at the same time using SIMD instructions is discussed in [185] and Chap. 3 and applied to SanssouciDB while focusing on bit encoding of integer values. The algorithm is based on 128-bit SIMD registers and the related instructions that are implemented in current processors, e.g. Intel SSE. Within a single step four 32-bit integer values are decompressed. After loading the 128-bit value into the SIMD register the relevant bytes containing the bit encoded values are identified. Following this step, the relevant bytes are copied to the appropriate location within the register using a specific command, the so-called Selective Shuffle. If the data represented by one compressed value is not aligned to the boundary of a byte, byte-alignment has to be established. This is done using a SIMD shift instruction that allows shifting four 32-bit integer values within the 128-bit register in parallel, applying variable shift amounts.

This SIMD approach can be leveraged to execute the following in-memory table scan operations with very short latency:

- **Vectorized value decompression**: during a scan operation, column values might have to be explicitly decompressed in order to eventually continue the query execution in operations like projections. In this case after completing the shift operation, the decompressed values are transferred to the main memory.
- **Vectorized predicate handling**: during a typical table scan operation, simple predicates like equal-value or value-range search queries must be executed. The SIMD decompression approach is adopted to realize a vectorized value search process without the need to decompress the actual column values. Rather than decompressing all the data and then start searching for the values, direct compressed comparison with a search condition (i.e. predicate) can be used instead.

5.2 Increasing Performance with Data Aging

We discovered that companies usually store data for a period of ten years in their operational system while only actively using about 20% of it, which is data created in the last two years. This motivates us to apply dynamic horizontal partitioning to reduce main memory usage and processing time by a factor five. We focus on improving query performance for active data to minimize response times for the most common queries. A definition of active and passive data will be given in the next section.

In general, horizontal partitioning is used to split tables with many rows into smaller tables in order to reduce table sizes, enabling faster operations and better scalability. A straightforward approach how to partition the data horizontally is to use attribute ranges, for example, to partition a customer table by zip code. Another strategy is hashing, which ensures that the resulting table are of similar size.

Horizontal partitioning was introduced in the 1970s [24]. It is related to fragmentation, allocation, replication, and query processing. Looking at these problems simultaneously is hardly possible due to the complexity of each of these problems. We only focus on fragmentation and highlight the resulting improvements for query processing in this section.

Horizontal partitioning was intensively discussed in the context of initial database design [20, 108]. This means that partitions are defined once and, depending on their properties, tuples are stored in a respective partition. If it is known how frequently certain data is accessed, this information can be exploited to partition the data to minimize total processing time. However, usually, no horizontal partitioning is leveraged to separate active and passive data.

The majority of enterprise applications follow the object-oriented programming model. Thus, they are comprised of a large number of business objects. In such a setting it is advisable to partition the data into active and passive data, as detailed information is available about the business object's lifecycle, context information, and access patterns. Active business objects should be separated from objects at the end of their lifecycle because they have different characteristics. In our approach to horizontal partitioning, we follow these considerations. In contrast to static horizontal partitioning, we move tuples between partitions at run-time depending on certain property changes, such as status field updates.

5.2.1 Active and Passive Data

Given our expertise in enterprise applications, we see that it is possible to divide transactional data into active data and passive data. Our approach is to move tuples related to a specific business object instance to another partition once their properties change to the effect that the respective business object instance becomes passive. The main use case is to move tuples of business object instances that are no longer actively needed out of the main partition. Active data is defined as data that is in use by one or more business processes and therefore likely to be updated or read in the future. This, for example, includes open invoices or recently closed invoices that are subject to operational reporting or shown in a dashboard. In contrast, we define passive data as data that is not used by any business process and is not updated or commonly read any more. Passive data is purely used for analytical and statistical purposes or in exceptional situations where former business transactions have to be manually analyzed.

The rules for the transition from active to passive have to be defined by the application and are inherently application-specific. If multiple applications work on the same business objects, the rules of all applications have to be evaluated. A business object instance cannot become passive until all applications allow the transition from active to passive. A rule of thumb is the following: once a business object instance is created, it is active. It remains active as long as any application works on it. As soon as it is certain that no application updates this business object instance any more, the business object instance and its related data can be aged and becomes passive. Passive data is 100% read-only and only used for analytics.

We have seen many companies that store data for the last ten years in their operational system. They inflate the DBMS with nearly 100% read-only data. By applying horizontal partitioning to passive data, the amount of data in the main store can be radically reduced. As processing times of database operations such as scans, joins, and aggregations are functions of the number of tuples in the database tables (see Fig. 5.3 for an example with two tables: leads and business partners), all operations that only rely on active data become faster. Another advantage we is that we can store passive data on less expensive hardware that nevertheless provides high performance especially for sequential read operations. These sequential reads are important for analytical applications such as planning or simulation that utilize historical data. As a result, we need less main memory to store all active data for in-memory processing

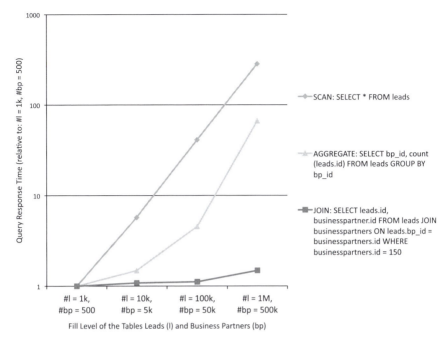

Fig. 5.3 Scalability of scan, join, and aggregate operations

while still achieving high performance for analytical queries, even if they comprise large amounts of passive data.

5.2.2 Implementation Considerations for an Aging Process

Aging is possible with SanssouciDB because it leverages the recompaction process (see Sect. 5.5). During this process, all tuples of a column are de-compressed, new tuples are added from the differential buffer, and the tuples are compressed again (see Sect. 3.3). The partitioning rules have to be defined for each business object separately by the business processes it is involved in. As enterprises strongly think in accounting periods and many business processes end at the year-end, we propose to have a partition pt for active data that is regularly accessed and updated. A partition p_{t-1} is used for data that is not updated anymore but used for analytics. The third partition p_{t-2+} stored passive data that is not regularly read anymore. This partition is kept for compliance reasons as well as infrequent investigations of historical data.

Tuple movement among the partitions can be transparent to the application layer—that is, nothing has to be changed in the application and the DBMS is in charge of processing the queries. To optimize query processing, we implement the behavior of SanssouciDB as follows: by default, only active data will be incorporated into query

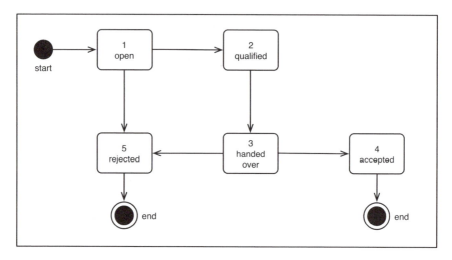

Fig. 5.4 Lifecycle of a lead

processing. If the application wants to access passive data, a respective flag has to be set within the SQL request. Given the fact that the application defines the aging behavior, it is aware of the data that is currently active and passive.

5.2.3 The Use Case for Horizontal Partitioning of Leads

To demonstrate our idea of run-time horizontal partitioning, we show in Fig. 5.4 the lifecycle of a lead as derived from an enterprise application. A lead is a contact that might be interested in one of the company's products. For example, when the marketing department of an organization gets business cards at a trade fair, collects addresses via a marketing campaign, or if a salesperson just hears that someone might be interested, these contacts are entered into an enterprise application.

The status of each lead is open in the beginning. The sales division decides whether it wants to follow that lead by changing the status to qualified or if the lead is uninteresting or invalid and can be rejected. Each qualified lead is handed over to a sales representative. The sales representative works on the lead by contacting the interested person or company. Finally, the salesperson decides whether the lead is irrelevant and can be rejected, or if there is real interest in a company's product and the lead is accepted and treated as a real sales opportunity.

Our simplified example database table for lead shall contain the following attributes: id, name, description, priority, created_at, origin, result_reason, status, and updated_at. The table including the aging metadata can be constructed as depicted in Listing 5.2.

Table 5.1 Example data

id	Status	Updated_at
1	Accepted	2001-12-15
2	Accepted	2002-04-12
3	Accepted	2003-06-08
4	Rejected	2004-07-30
5	Accepted	2005-02-08
6	Accepted	2006-04-19
7	Rejected	2007-03-28
8	Rejected	2008-01-03
9	Handed over	2009-04-22
10	Open	2010-06-22

Listing 5.2 (SQL Statement to Create our Simplified Lead Database Table)

```
CREATE TABLE lead(
    id INT,
    name VARCHAR (100),
    description TEXT,
    priority INT,
    created_at DATETIME,
    origin INT,
    result_reason INT,
    status INT,
    updated_at DATETIME),
AGING := (
    status == 4|5 &&
    updated_at < SUBSTR_YEAR(Time.now, 1)
);
```

To exemplify how horizontal partitioning is applied, we use the following data in Table 5.1.

From a business perspective, leads are only relevant if their status is neither 4 (accepted) nor 5 (rejected): the lead is not at the end of its lifecycle. Nevertheless, leads are still active if their status is 4 or 5, and they were updated less than one year ago. Accordingly, we define two partitions for the leads' database table: the first partition pt contains all active tuples, meaning that their status is neither accepted nor rejected or they are updated less than one year ago. The second partition p_{t-1+} contains all passive tuples where the status is either accepted or rejected and they are last updated more than a year ago.

With regard to our example data, this means that it is partitioned as follows (Tables 5.2, 5.3).

In this simple example the main store (see Sect. 5.5) has shrunken to 20% of its original size. In a customer-data analysis, we found that less than 1% of sales orders are updated after the year of their creation. This encourages us in our approach

Table 5.2 Partition p_t for active data

id	Status	Updated_at
9	Handed over	2009-04-22
10	Open	2010-06-22

Table 5.3 Partition p_{t-1} for passive data

id	Status	Updated_at
1	Accepted	2001-12-15
2	Accepted	2002-04-12
3	Accepted	2003-06-08
4	Rejected	2004-07-30
5	Accepted	2005-02-08
6	Accepted	2006-04-19
7	Rejected	2007-03-28
8	Rejected	2008-01-03

because we can move sales orders that are at the end of their lifecycle to another partition to increase query performance and reduce memory consumption significantly.

To summarize this section, we conclude that horizontal partitioning is applied to separate active and passive data. The definition of active and passive is given by the applications for each business object it uses. The goal is to optimize overall processing time by decreasing query processing time for the majority of queries that only deal with active data. Other positive effects are that less space is needed to store all relevant data of an enterprise in main memory and system performance is accelerated.

5.3 Efficient Retrieval of Business Objects

Business applications expect that data is organized in certain (typically object-oriented) format. A business object is an entity that is significant to a business, for example, an employee, a business partner, or a sales order. From a technical aspect, business objects encapsulate semantically related data together with the business logic and the rules required to manipulate the data. In contrast to the general notion of object in an object-oriented sense, the term business object is used here to describe the type and not the instance level. Consequently, the business object Sales Order describes the category of all sales orders and not a single instance of a sales order. In SanssouciDB, a business object is represented by a tree whose vertices are business object nodes. A simplified version a business object is shown in Fig. 5.5. Each business object node represents a set of semantically related attributes. For the sake of simplicity, we assume that every business object node corresponds to a database table in which every node instance corresponds to a tuple. Every node instance that is not an instance of the root node stores the primary key of its parent instance as a foreign key. This information is sufficient to reconstruct an instance of a business object by starting at the root node instance and recursively reading all child nodes. Some nodes may contain references to other business objects. This representation allows modeling of business objects and business processes.

5.3.1 Retrieving Business Data from a Database

If an application requests a complete instance of a business object, the result of this operation is a nested relation, a set of relations, or a sparse relation based on the union of all attributes that make up the business object, along with an extra attribute that denotes the business object node type. The rows in the sparse relation correspond to the instances of the business object nodes. Each tuple in the relation contains only values for the attributes present in that particular node. The implementation of this operation needs information about the business object model to be incorporated into the physical data model.

A straightforward implementation of this operation uses the recursive relation to read the children of every node instance iteratively level by level and, within each level, node by node. Disjoint sub-trees of the business object structure can be processed in parallel. In the example shown in Fig. 5.5, the children of the root can be read in two threads for nodes Header and Item, but the instances of node ScheduleLine can only be read when the corresponding instances of the node Item are known. Of course, it is possible to pipeline identified instances of the node Item to start searching for schedule lines. But in this case a more intelligent query engine is required, and the performance is still not optimal. Our observation shows that in many cases only a small fraction of the nodes has instances for a given business object instance. The naïve implementation requires searching in many tables that are not relevant.

Since business objects can have dozens to hundreds of nodes and each table can feature many attributes, retrieving of a complete business object instance can impose a performance issue.

5.3.2 Object Data Guide

SanssouciDB improves the retrieving performance of a business object by adding some redundancy to the physical data model. This redundancy represents a join index for querying sparse tree-shaped data, which is called object data guide and includes two aspects:

- In addition to the parent instance, every node instance can contain the link to the corresponding root instance (ROOT_ID). Using this additional attribute, it is

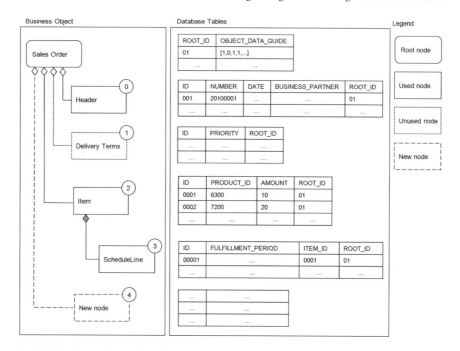

Fig. 5.5 Business object sales order

possible to retrieve all nodes in parallel instead of waiting for the information from the parent level.

- Each node type in a business object can be numbered, and then for every root instance a bit vector (OBJECT_DATA_GUIDE) is stored, whose bit at position *i* indicates if an instance of node number *i* exists for this root instance. Using this bit vector, a table only needs to be checked if the corresponding bit is set. In Fig. 5.5, the table numbers are marked in circles. Note that the business object node marked as unused corresponds to the concrete instance and not to the business object node itself. This formatting is used here due to readability purpose. In reality, only concrete instances can be marked as used or unused. The responsibility for the maintenance of this additional information is taken by the physical storage engine. The bit vector of the object data guide is an aggregate from the physical point of view. Therefore, concurrent modifications of the bit vector must be managed.

The bit vector and the ROOT_ID allow querying of all tables in parallel. Query parallelization is limited only by the number of available cores. Although object data guide slightly increases query scheduling overhead, the total performance can be significantly improved.

Fig. 5.6 Complexity tension
between execution on the
application server versus
execution in the IMDB

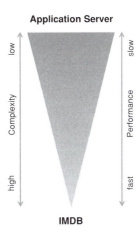

5.4 Efficient Execution of Business Functions

With the increasing capabilities of in-memory databases it becomes viable to move
business logic closer to the actual execution layer of the database. In this section
we discuss the advantages and disadvantages of this procedure and will give a few
examples to illustrate the general applicability.

5.4.1 Separating Business Functions from Application Functions

Traditionally enterprise applications follow a three tier architectural model, where the
application logic is encoded in the application server and the persistence is handled
by the database server. Typically the task of the persistence layer is to handle the write
load and to execute the queries requested by the application server. Interpreting the
queries or understanding business logic is not part of the original contract between
the layers.

With in-memory databases, the database layer can execute complex queries faster
thus allowing for new scenarios and applications. In this case, the persistence layer
needs to be able to execute new functionality that resembles business logic and
semantics. This is to avoid costly data transfer between the application layer and the
database layer.

Figure 5.6 shows the tension between the layers and illustrates the different factors that have an influence on the traditional picture. The right arrow in the figure
illustrates the execution performance of a piece of application logic. The closer
the application logic is to the in-memory database system, the faster the execution
becomes. However, this is opposed by the degree of complexity such an optimization imposes. Application logic put close to the database typically requires a different

programming language or programming model such as stored procedures or database operators. Compared to the object-oriented approaches of application servers that try to hide complexity and let the user focus on the business semantics, these stored procedures or database operators are focused on performance and trade full-flexibility for usability making them more difficult to use.

- *Reuse*—is the functionality customized for one specific use-case or is it reusable by different applications?
- *Business centric*—is the functionality centered around the business processes built with it?
- *Data centric*—does the business function consume and produce large amounts of data?
- *Calculable*—the logic expressed in a business function library should express business context using mathematical expressions.

It is important to differentiate business functions from classical stored procedures. In contrast to stored procedures, business functions should be free of side-effects and stateless. Stored procedures, however, allow one to bundle more complex tasks together and resemble a more process-like behavior. From a complexity perspective, business functions on the lowest level are simple mathematical functions ($+/-$), followed by aggregation functions (`sum`, `avg`). Business functions are defined following this definition:

Definition 1 A Business Function applies mathematical and statistical functions to enterprise data following business rules, itself being stateless and free of side-effects.

Since the complexity of business functions is higher compared to the operations mentioned above it is crucial that the information about their implementation and availability is documented and easily accessible by developers. One can describe business functions as the extended standard library of programming for enterprise applications.

Simple examples of business functions include clustering algorithms for customers and products or forecasting algorithms based on current demand.

5.4.2 Comparing Business Functions

In the following sections we will introduce a few examples that show what kind of business functionality should and can be moved closer to the database.

5.4.2.1 Calculating Growth

When using in-memory databases for analytical and planning applications, or even sometimes in day-to-day operational analytics, it is often desirable to calculate growth

Table 5.4 Example for linear and compound growth

	Jan	Feb	Mar	Apr	May	Jun	Jul	Aug	Sep
Base	800	800	900	1,000					
Rate	–	–	–	0.1	0.1	0.1	0.1	0.1	0.1
Linear growth	–	–	–	1,100	1,200	1,300	1,400	1,500	1,600
Compound growth	–	–	–	1,100	1,210	1,331	1464.10	1610.51	1771.56

over a time series. A very simple example is the calculation of the expected sales for the current year. The input values for this operation are the time period data and a few other parameters, for example, the growth rate, growth algorithm and the number of periods to grow.

In our case the business function library will now distinguish between two different growth algorithms: linear and compound growth. The linear growth can be defined using Eq. (5.1) and the compound growth can be described as shown in Eq. (5.2).

$$Growth Result Linear_n = Growth Result_{n-1} + (Base_i * rate_{n-1}) \qquad (5.1)$$

$$Grow Result Compound_n = Growth Result_{n-1} + (Growth Result_{n-1} * rate_{n-1}) \qquad (5.2)$$

The growth rate for both formulas is a float value between 0.0 and 1.0. When all input parameters are present we can call the business function directly using SQL, while the original library function would be implemented highly efficiently using C or even Assembler.

An example output for the growth function is shown in Table 5.4. The table shows input data for the period from January to April and expects to calculate the growth for the period until September, based on the initial information from April.

To summarize, calculating the growth is a very good candidate for inclusion in a business function library because, for larger time series, the operation consumes and produces a large amount of data. That means it is highly reusable for different scenarios and can be directly assigned to a business use case.

5.4.2.2 Counter-Example: Converting or Resizing Images

Even though graphics processing requires a comparatively high amount of resources, moving image processing logic violates the *business centric* principle, because image manipulation is not a required component for the business functionality.

As we have seen, a business function library exposes a set of reusable, business and data centric methods that are consumed directly on the database level within the business application context. However, not all functionality is equally optimal to be represented in such a business function library, only the necessary functions should be pushed down.

5.5 Handling Data Changes in Read-Optimized Databases

The data layout of SanssouciDB is based on the observed workload (see Sect. 4.4). Transactional data mainly accessed with analytic-style queries is organized in a column-oriented layout. Such read-optimized data structures are required to process a mixed workload that consists of both set processing and singe object access in an efficient manner. As pointed out in Sect. 4.3, read performance of column-oriented data is increased by lightweight compression schemes improving cache awareness especially of scan and aggregation operations.

Unfortunately, compression induces the problem that changes on a compressed structure are prohibitively slow, because—depending on the employed compression algorithm—much more data than actually manipulated must be de- and recompressed. To address this issue, every relation stored in a column-oriented layout is associated with an additional write-optimized insert-only structure called differential buffer. This structure is sometimes referred to as differential store. All inserts, updates, and delete operations are performed on the differential buffer.

The read-optimized main part of the data, called the main store, is not touched by any data modifying operation. The differential buffer in conjunction with the compressed main store represents the current state of the data. The differential buffer also facilitates a column-oriented, but less compressed data structure that can be updated efficiently. To address the disadvantages of data compression, write-performance is traded for query performance and memory consumption by using the differential buffer as an intermediate storage for multiple modifications. Figure 5.7 depicts the overall idea of the differential buffer. To derive a consistent view, read operations must access the main store as well as the differential buffer, while data modifying operations manipulate the differential buffer only.

The differential buffer grows with every write operation. This decreases read performance, since an increased part of the data has to be accessed via a non read-optimized data structure, the differential buffer. To compensate for this effect, the differential buffer and the main store are merged from time to time within the so-called merge process. The merge process is carried out per table and includes computing a new compression scheme for a new combined main store, replacing the old main store. Deriving a new compression scheme includes decompression of all the tuples encoded in the main store and differential buffer as well as compression of the combined data, which is expensive in terms of memory and computational effort. Implementing an efficient merge strategy and scheduling the point in time to execute the merge is therefore crucial so as not to degrade the performance of the DBMS.

The approach of using a differential buffer for write operations and merging the data of the differential buffer with the main store is also described in [97] and [98]. Related work in the context of disk-based databases can be found in [5] and [164]. In this section we will describe the concept and implementation of the differential buffer in detail and show how OLTP queries operate on the differential buffer. The second part of this section examines different implementations of the merge process.

Fig. 5.7 The buffer concept
with regard to the database
operations

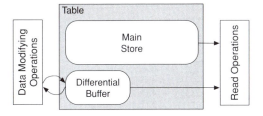

5.5.1 The Impact on SanssouciDB

In SanssouciDB the default compression scheme is dictionary encoding, where a
dictionary is used to map each distinct value of a column to a shorter, so-called,
value ID (see Sect. 4.3.1). Value IDs are used to represent the actual values of records.
Each column of the main store in SanssouciDB consists of two arrays as shown in
Fig. 5.8. The dictionary array contains the values in sorted order, while the position
of a value defines its value ID. This order allows for binary search on the dictionary
vector, if the value ID for a given value is searched. For example, the value Beta is
mapped to value ID 1, based on its position in the dictionary. Each value ID is bit
compressed, using as many bits as needed for fixed length encoding. Since in our
example the main store contains only four distinct values, two bits are sufficient to
decode all value IDs. Inserting a new value, for example Echo into the main store,
would require to re-encode all values, since now two bits are not sufficient anymore
to encode all required value IDs. Additionally, inserting a new distinct value that
changes the sort order requires a re-encoding of the complete column. In practice, a
single write operation might cause the en-coding and de-coding of millions of values,
this illustrates again that modifying the main store directly is prohibitive.

The following examples point out this fact:

- *Update without new distinct value*: as the new value of the updated attribute of a
  tuple is already in the dictionary this leads to a replacement of the old value ID
  with the new value ID read from the dictionary.
- *Update with new distinct value*: the new value is not in the dictionary. Therefore,
  the new value extends the dictionary while it may also change the sort order of the
  dictionary and force more value IDs to be changed than only the selected one. In
  the event that the cardinality of the distinct values reaches $2n + 1$, with n being the
  old cardinality of distinct values, the complete attribute vector must be recalculated
  regardless of any sort order change in the dictionary.
- *Insert without new distinct value*: a new tuple is appended to the table using an
  existing value ID from the dictionary. This is essentially the same as an update
  without new value.
- *Insert with new distinct value*: inserts to the table occur with a new value that is
  not yet in the dictionary. This event has the same complexity as the case Update
  with new distinct value described above.

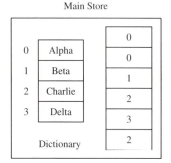
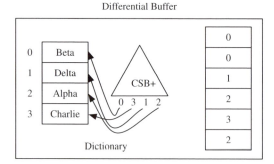

Fig. 5.8 Data structures in SanssouciDB

- *Delete without value change*: since the offset of a specific value in the attribute vector is used as the implicit row number to enable tuple reconstruction of the decomposed relation, a delete would force a complete lock of the table until all queries are finished.

Similar to the main store, the differential buffer consists of a dictionary and a vector of value IDs. To improve write performance of the differential buffer, efficient insertion of new unique values into the dictionary is required. In SanssouciDB this is achieved by using the CSB+ tree [145]. The CSB+ tree is mainly used as an index to look up value IDs for a given value while new values can be inserted easily. Figure 5.8 depicts the basic structure of the main store and differential buffer if dictionary encoding is applied.

Since the differential buffer is an insert-only structure, an insert as well as an update operation creates a new row in the differential buffer. Each record inserted into the differential buffer has an additional system attribute Invalidated Row, which indicates if this entry invalidates a tuple in the main store. In case the new tuple invalidates an existing tuple of the main store, the row ID of this tuple in the main store is saved in this field. Delete operations are handled similar using a special tuple type in the differential buffer.

5.5.2 The Merge Process

Merging the write-optimized differential buffer and the read-optimized main store at some point in time is required to maintain the read performance. The reasons for merging are twofold. On the one hand, merging the data of the differential buffer into the main store decreases the memory consumption since better compression techniques can be applied. On the other hand, merging the differential buffer allows improved read performance due to the ordered value dictionary of the read-optimized main store. In an enterprise context, the requirements of the merge process are that it runs asynchronously, has as little impact as possible on all other operations, and does

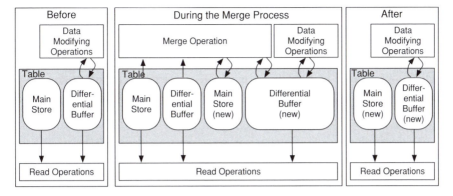

Fig. 5.9 The concept of the online merge

not delay any OLTP or OLAP queries. To achieve this goal, the merge process creates a copy of the main store and a new differential buffer to reduce the time where locks on the data are required. In this online merge concept, the table is available for read and write operations during the merge process. We will describe the merge algorithm and acquired locks in detail later. During the merge, the process consumes additional resources (CPU and main memory) that have to be considered during system sizing and scheduling.

When using a differential buffer, the update, insert, and delete performance of the database is limited by two factors: (a) the insert rate into the write-optimized structure and (b) the speed by which the system can merge the accumulated updates into the read-optimized main store. It is important to mention that all update, insert, and delete operations are captured as technical inserts into the differential buffer while a dedicated valid tuple vector per table and store ensures consistency. By introducing this separate differential buffer the read performance is decreased depending on the number of tuples in the differential buffer. Join operations are especially slowed down since results have to be materialized. Consequently, the merge process has to be executed if the performance impact is too large.

The merge process is triggered by one of the following events:

- The number of tuples in the differential buffer for a table exceeds a defined threshold.
- The memory consumption of the differential buffer exceeds a specified limit.
- The differential buffer log for a columnar table exceeds the defined limit.
- The merge process is triggered explicitly by a specific SQL command.

To enable the execution of queries during a running merge operation, the concept of an asynchronous merge process that does not delay transaction processing is required. Figure 5.9 illustrates this concept. By introducing a second differential buffer during the merge phase, data changes on the table that is currently merged can still be applied. Consequently, read operations have to access both differential buffers to get the current state. In order to maintain consistency, the merge process requires a lock at the beginning and at the end of the process while switching the stores and apply necessary data such as valid tuple modifications, which have occurred during merge runtime. Open transactions are not affected by the merge, since their changes are copied from the old into the new differential buffer and can be processed in parallel to the merge process. To finish the merge operation, the old main store is replaced with newly created one. Within this last step of the merge process, the new main store is snapshotted to a permanent store and defines a new starting point for log replay in case of failures (see Sect. 5.8).

The merge process consists of three phases: (i) prepare merge, (ii) attribute merge, and (iii) commit merge, while (ii) is carried out for each attribute of the table. The prepare merge phase locks the differential buffer and main store and creates a new empty differential buffer store for all new inserts, updates, and deletes that occur during the merge process. Additionally, the current valid tuple information of the old differential buffer and main store are copied because these may be changed by concurrent updates or deletes applied during the merge, which may affect tuples involved in this process.

The attribute merge phase as outlined in Fig. 5.10 consists of two steps, which are run sequentially to reduce the memory overhead during the merge process. If enough main memory is available, these two steps could be run in parallel. In the first step of the attribute merge operation, the differential buffer and main store dictionaries are combined into one sorted result dictionary. In addition, a value mapping is created as an auxiliary mapping structure to map the positions from the old dictionary to the new dictionary for the differential buffer and main store. Iterating simultaneously over the input dictionaries creates the new dictionary while the CSB+ tree of the differential buffer dictionary is used to extract a sorted list of the differential buffer dictionary that is needed for the algorithm. The current values of both dictionaries are compared in every iteration and the smaller value is added to the result and the corresponding iterator is incremented. In the event that both values are equal, the value is only added once and both iterators are incremented. In order to be able to later update the attribute vector to the newly created dictionary, every time a value is added to the merged dictionary the mapping information from the corresponding old dictionary to the merged one is also added to the mapping vector, which is an auxiliary structure for the second step. If the sizes of the dictionaries differ, the remaining values are directly added to the result since both input directories are sorted internally.

In the second step of the attribute merge, the values from the two attribute vectors are copied into a new combined attribute vector. During the copy operation the mapping the new dictionaries is applied and the new value ID is copied to the vector.

The commit merge phase starts by acquiring a write lock of the table. This ensures that all running queries are finished prior to the switch to the new main store including the updated value IDs. Then, the valid tuple vector copied in the prepare phase is applied to the actual vector to mark invalidated rows. As the last step the new main store replaces the original differential buffer and main store and the latter ones are unloaded from memory.

Although full attribute scans can be executed with high performance in the column-oriented storage, it is necessary to maintain a group index for foreign key

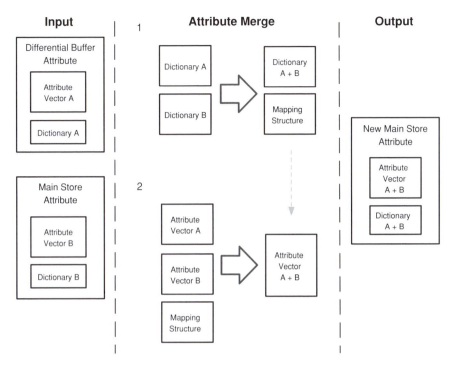

Fig. 5.10 The attribute merge

columns or often used predicate attributes to increase the performance for join and aggregation operations. Typical examples are retrieving all movements for a given material, aggregating the sum of all sales orders for a given customer. In contrast to disk-based databases, the group index does not only store the page number but the explicit offset of the tuple in the table. Since this offset may change during the merge process due to inserts, updates or deletes, the group indices have to be either rebuilt in case an index was not previously loaded or updated with the position offsets of the newly inserted values. These maintenance operations on the index have to be considered when scheduling the merge in addition to the pure data level operations on the table.

Figure 5.11 shows the result of the merge process for a simple example. The new attribute vector holds all tuples of the original main store, as well as the differential buffer. Note that the new dictionary includes all values from the main and differential buffer and is sorted to allow for binary search and range queries that incorporate late materialization query execution strategies.

During the merge phase, the complete new main store is kept inside main memory. At this point, twice the size of the original main store plus differential buffer is required in main memory to execute the proposed merge process. Since in enterprise applications tables consist of millions of tuples while having hundreds of attributes, this can lead to a huge overhead since at least twice the largest table size has to be

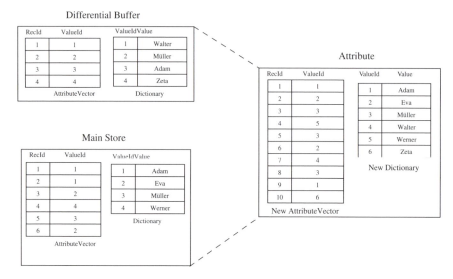

Fig. 5.11 The attribute in the main store after merge process

available in memory to allow the merge process to run. For example, in the financial accounting table of a large consumer products company contains about 250 million line items with 300 attributes. The uncompressed size with variable length fields of the table is about 250 GB and can be compressed with bit compressed dictionary encoding to 20 GB [99]. However, in order to run a merge process at least 40 GB of main memory are necessary. In the subsequent section a modification of the algorithm to decrease the overall additional memory consumption is described [98].

5.5.3 Improving Performance with Single Column Merge

The goal of the modified merge process called Single Column Merge is to reduce the size of memory consumption throughout the merge process. By merging single columns independently from the differential buffer into the main store, the algorithm reduces the additional memory consumption to the size in memory of the largest column. To use this technique the insert-only strategy has to be used; otherwise tuples would be physically deleted which could lead to inconsistent surrogate/row identifiers if merged columns are applied independently.

In the merge process described in the previous section, the merge result for single columns is calculated independently in the respective attribute merge phases. The merge result is kept in main memory until all attributes are merged to ensure an instant switch to the new main store in the commit merge phase. The basic idea of single column merge is to switch to an updated main store after every attribute has been merged while maintaining a consistent view on the data. Consequently, a partial

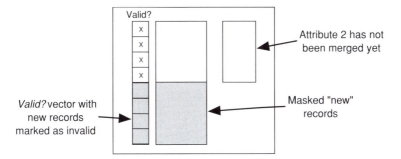

Fig. 5.12 The main store during the single column merge

hiding of merge results is needed because switching already merged columns leads to a problem: some attributes are already merged while others are not. Those finished attributes typically have a longer attribute vector since new rows could have been inserted into the differential buffer. And as this buffer is not updated throughout the merge process value entries for newly created rows are duplicated in the updated main store and original differential buffer. To resolve this issue all newly created rows are marked as invalid until all columns are merged as shown in Fig. 5.12.

Due to a possible change of the sort order in the dictionary a remapping of old value IDs is necessary. After one attribute is merged, its state differs from the rest of the index that has yet to be merged. Some values potentially have new value IDs if the merge process has changed the value IDs. Incoming queries might still rely on old value IDs, for example, if they have been cached by queries started prior to the merge process. To avoid locking of the table for each attribute, a mapping table from the old value IDs to the new ones is provided throughout the merge process until all attributes are merged into the main store. This mapping table from old to new values is created in the attribute merge phase when merging the dictionaries of differential buffer and main store. Figure 5.13 shows an example for a remapped lookup of the cached old value IDs 1 and 4.

The following modifications of the traditional merge process have been applied to implement the single column merge:

- *Prepare merge phase*: the valid tuple vector of the main store has to be enlarged by the number of rows that are currently in the differential buffer. This is required to hide the newly created merge results in the main store until all attributes of the table are merged. The newly created valid tuple entries are initialized with false to deactivate those rows.
- *Attribute merge phase*: for each attribute the following changes have to be made: retain the mapping tables from old to new value IDs in memory. To have a consistent view of the data, these tables have to be provided to functions in the query executor that might be called while a merge is running. Switch the attribute data structure of the old main store to the merge result right after merging the attribute.

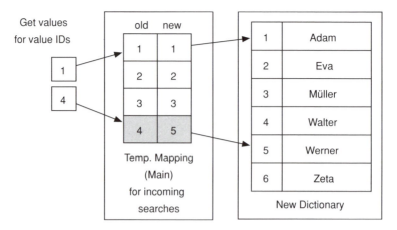

Fig. 5.13 Remapping during the single column merge

- *Commit merge phase*: activate the newly merged rows by setting the valid tuple vector entries to true. Unload mapping tables from old to new value IDs after the lock on the table is acquired.

   With regard to the memory consumption, applying the single column merge eliminates the need to additionally hold the newly created main store of the size of the original main store and differential buffer in main memory. As only one attribute is merged at a time, the additional amount of main memory needed for the merge process is the size of the attribute data structure currently merged plus the size of the mapping tables from old value IDs to new value IDs for the dictionaries. Assuming that the main store is significantly larger in size than the differential buffer, the overall additional memory consumption for the merge process is driven by the size of the largest data structure of all attributes.

   To test how large the savings in additional memory consumption are, we compared the traditional merge process and the single column merge using live customer data. The two major tables in the database consist of 28 million rows with 310 columns and 11 million rows with 111 columns. The main memory usage during the test is shown in Fig. 5.14. The graph shows the additional memory consumption during a merge process for both merge strategies. The column that consumes the most memory can be seen in both test series. The main memory usage during the single column merge clearly peaks at around the size of the largest column, as opposed to the steadily increasing memory usage during the traditional merge.

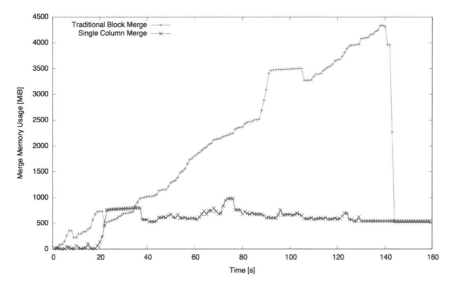

Fig. 5.14 Memory consumption

5.6 Append, Never Delete, to Keep the History Complete

Data can change over time and these changes have to be persisted on the database level. Instead of updating existing data tuples, we propose an insert-only approach that appends new tuples to the data storage. The technical definition of insert-only data storage is a database that does not allow applications to perform updates or deletions on physically stored tuples of data [68]. Changes to existing data (logical updates) are handled by the database using explicit valid/invalid flags or timestamps. In earlier research this is called no-overwrite data management [166].

There are manifold reasons for using this technique: first, it allows a system to keep track of a data tuple's history, which is legally required for financial applications in many countries. Secondly, it enables a company to analyze its development over time and derive better business decisions for the future. Additionally, insert-only can simplify the implementation of Multiversion Concurrency Control (MVCC).

In the following we illustrate implementation strategies of the insert-only approach, and we discuss Concurrency Control (CC) techniques in the context of insert-only. Finally, the impact on enterprise applications is examined.

5.6.1 Insert-Only Implementation Strategies

In principle, there are two ways to implement the insert-only concept at the database level: point representation, which stores only the delta of the new tuple to the old one with a single timestamp, and interval representation, which stores the full tuple including additional information that is used to determine the time span of its validity.

5.6.1.1 Point Representation

In the case of an update, this implementation strategy only stores the delta of the changed value at a specific point in time. This significantly reduces the amount of data required for each insert operation. In addition, not all columns have to be touched for this operation; that is, only the changed attribute(s) and the timestamp of insertion. To modify data, not the whole row needs to be touched, because all non-changed attributes can be filled with a default value for not changed. When retrieving the tuple, however, all older versions have to be read to restore the full tuple and to determine the current valid value.

This implementation strategy is best suited for OLTP applications, since the critical operation is the insertion of new or changed data tuples. The insert operation must be performed quickly while the need for further post-processing of inserted data is eliminated. The main drawback of this method is the requirement to touch and aggregate all older versions of the data to reconstruct the valid version for a specific time. This becomes more expensive if there are a lot of changes within the data.

5.6.1.2 Interval Representation

The second way of implementing the insert-only approach is to store each new tuple explicitly by using an interval-based representation with timestamps. When performing a logical update to a tuple, it has to be determined which of the tuples is the most recent. This can be achieved by storing an additional insertion timestamp or an automatically incremented identifier. Thus, the database system has to ensure that just the most recent tuple is returned at query time. In detail, the interval representation can be implemented by storing two timestamps for each row: `valid from` and `valid to`. When a new tuple is inserted only the `valid from` timestamp is set and the `valid to` timestamp is left open. As soon as an update to a tuple is received, the valid to field of the old tuple is set to the current timestamp, and the new tuple is saved like a new insertion. Although the old tuple has to be touched, the advantage is improved search performance. Since the validity interval is already stored for each tuple, it is simple to find the correct matching tuple without touching all old data. These timestamps also enable applications to time travel and retrieve data that was valid at a certain point in time by filtering the data based on the valid from and valid to timestamps. Time travel is also possible with the point representation; yet, it is more expensive. The interval representation is best suited for applications where reading is the most dominant data access pattern, which is commonly found in OLAP applications. To achieve the best possible performance it is reasonable to

retrieve the desired value by the read operation directly, that is, without the need for
any preprocessing.

5.6.2 Minimizing Locking Through Insert-Only

To leverage the multi-core architectures available today, database queries should be
parallelized. This necessitates the use of adequate CC techniques to ensure transactional isolation. We briefly introduce common CC concepts, outline how insert-only
can simplify their implementation, and exemplify this with SanssouciDB.

5.6.2.1 Concurrency Control

There are two principles that guarantee a performance advantage for parallel execution while using CC [65]. On the one hand, the results of serial and parallel execution
must not differ. Common problems are dirty reads, lost updates, and incorrect summary [48]. On the other hand, parallel execution should never be slower than its
equivalent serial execution. For instance, parallel execution of multiple transactions
may result in deadlocks and high blocking times (in the case, locks are used), which
do not exist when transactions are executed serially. The complex CC problem is
typically solved by shielding the execution of parallel transactions from each other
in two ways: either by isolating transactions with locks or by using optimistic CC
techniques.

5.6.2.2 Locking

Data changes can be isolated from each other by using a special kind of resource
allocation that ensures serial execution. One way of achieving this is via the use of
locks that allow the lock owner to manipulate the locked resource exclusively. Data
manipulation is only started if the system can ensure that the required resources are
available, that is, not locked.

Locking has the disadvantage that once a tuple, row, or table is locked by a
transaction T, any concurrently executed transaction T_i trying to access this resource
is blocked until the desired lock is released. The overhead for this locking approach
reduces the transactional throughput, and potential waits for resources can reduce
the response time, which is at odds with the second principle of CC. Some modern
database systems try to prevent the need for locking completely by introducing a
single-threaded writer process, which performs all writing operations in sequence
and multiple reading threads in parallel.

Locking support at the database level is a technique that has been around for
many years. Moving the question of locking up to the application level may result in
code reduction in the database and in additional flexibility in the application level.

Application semantics can be used to optimize locking granularity, lock duration, and the need for locking, which might even result in lock elimination.

5.6.2.3 Optimistic Concurrency Control

With optimistic CC, the current state of the complete data set is stored in a virtual snapshot and all manipulations work on a set of copied data, valid from a specific point in time. Data manipulation is started regardless of whether other transactions are performing concurrent manipulations. Conflict checks are done shortly before the transaction finishes by determining whether the performed manipulations interfere with other changes performed in parallel by other transactions. If no conflict is detected, all changes are made persistent. If, on the other hand, a conflict is detected, the transaction is aborted and restarted.

5.6.2.4 Multiversion Concurrency Control

With MVCC, transactions that need to update data actually insert new versions into the database, and concurrent transactions still see a consistent state based on previous versions [48]. This way, read transactions do not require to lock resources and therefore do not block other transactions. Read operations read the version of a tuple valid at the beginning of a transaction, but write operations have to write the latest version of a tuple. While the writes still have to be synchronized, which can be implemented with locks, the reads do not have to be synchronized and thus increase transaction parallelism.

Implementations of MVCC consist of the following steps: first, each data value is tagged with a strict monotonically increasing version number or timestamp. After starting a transaction, the current version number vT for transaction T is determined, because its value defines the validity threshold for data values that were committed before the start of T. During the execution of T, only data values with a version number less than or equivalent to vT, that is, data tuples that have been committed before the start of transaction T, are visible to this transaction. Data values with a higher version number indicated that they are more recent and became valid after the start of T.

Since MVCC may create multiple versions of a data tuple, it necessitates the maintenance of old versions. This can be handled by implementing a garbage collection mechanism removing all temporary versions once there is no longer any running transaction that needs to access that state of the database. Insert-only on the other hand simplifies the implementation of MVCC by keeping all old versions.

5.6.2.5 Concurrency Control in SanssouciDB

In the following, we briefly describe the implementation of MVCC and its integration with the insert-only approach of SanssouciDB. For a more detailed description, we refer the interested reader to [136].

To provide a consistent view for a given transaction, information about the state of other transactions at the time the transaction started is required. Such information is maintained by a central transaction manager that keeps track of the order in which write transactions have been started and committed as well as of their state, which can be open, aborted, or committed. At transaction start, a transaction contacts the transaction manager and receives a so-called transaction token that contains its transaction ID, which reflects the order of transactions, and all information required to determine visibility of data tuples for this transaction. The token contains information of which transactions are committed and thus must be visible. In the same way, changes made by transactions started after the transaction can be identified and must be hidden in order to ensure isolation. Since the chosen insert-only approach explicitly requires SanssouciDB to keep all versions of a data tuple, these versions can be directly used by the MVCC protocol and thus does not require any additional maintenance of versions.

The transaction token concept makes it easy to implement different transaction isolation levels: for transaction level snapshot isolation, a transaction token is provided at transaction start and used for all statements in this transaction. For read committed isolation level (namely, statement level isolation) a new updated transaction token is provided for each statement.

5.6.2.6 Time Travel Queries

Time travel queries enable applications to retrieve values of a data tuple at different points in time. For a time travel query, the transaction manager reconstructs a transaction token that represents a historical point in time. With this transaction token, the statement runs against the consistent view on the database as it was at the specified point in time. To reconstruct historical transaction tokens, the transaction manager maintains a persisted transaction token history table that is written whenever the central transaction token is updated.

Listing 5.3 (Storage of Historical Versions)

```
UPDATE T1 SET Size='Large'  WHERE ID='454273'
```

To process time travel queries, SanssouciDB maintains the additional internal system columns Valid from and Valid to that indicate the validity of a row in a table, as illustrated in Fig. 5.15. The values in these system columns are Commit IDs (CIDs) that can also be used as timestamps in the event that there is a mapping between CIDs and the corresponding timestamps in a separate table. The Valid from value is the CID of the transaction that created the tuple. For data that is still

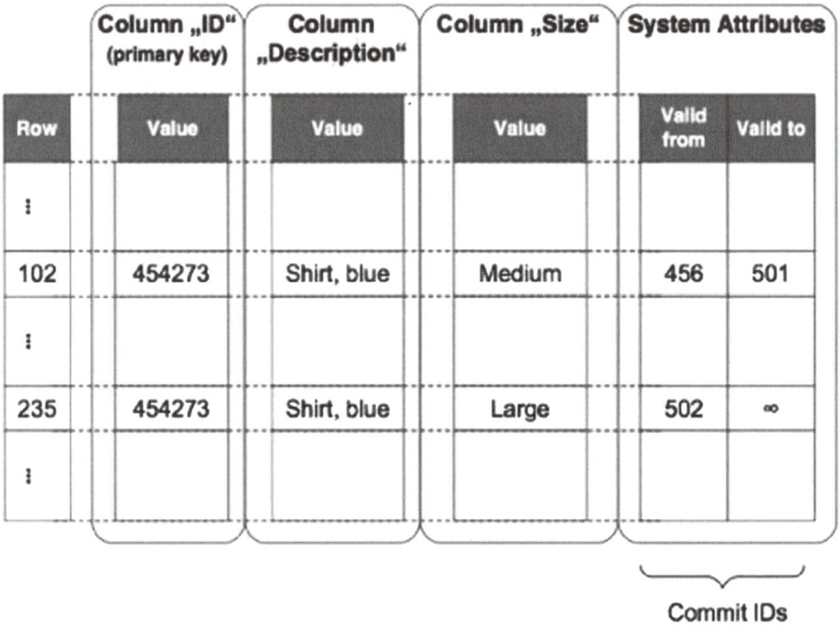

	Column „ID" (primary key)	Column „Description"	Column „Size"	System Attributes	
Row	Value	Value	Value	Valid from	Valid to
⋮					
102	454273	Shirt, blue	Medium	456	501
⋮					
235	454273	Shirt, blue	Large	502	∞
⋮					

Commit IDs

Fig. 5.15 Storage of historical versions

current, the Valid to value is not set, which means that it is infinite. If the Valid to value is set, the row contains an old version of the tuple.

To determine whether a row can be seen by a query, the maximum CID (maxCID) in the transaction token is evaluated, that is, the ID of the transaction that has committed most recently. The rows with Valid from ≤ maxCID ≤ Valid to interval are visible with the given transaction token. When a tuple is updated (see query in Listing 5.3), a new version is created and the previous version becomes historic by setting the Valid to timestamp. When a tuple is deleted, the most recent version before the deletion is made a historical version by setting the Valid to attribute accordingly. No specific deleted version is required.

5.6.3 The Impact on Enterprise Applications

In this section, we analyze the implications of the insert-only approach for enterprise applications. In particular, we first analyze the types of updates found in SAP's financial applications. Further, we illustrate new use cases such as time travel that are now possible with the insert-only approach, and we elaborate on the feasibility of this approach by including an analysis of customer data and its update behavior.

5.6.3.1 Types of Updates

By analyzing the change logs of four different SAP customers, we found that updates can be categorized into three major types [135]:

- *Aggregate update*: the attributes are accumulated values as part of materialized views, typically between one and five updates for each accounting line item.
- *Status update*: binary change of a status variable.
- *Value update*: the value of an attribute changes by replacement.

5.6.3.2 Aggregate Updates

Most of the updates taking place in financial applications apply to totals depending on the structure of the organization and the application. Totals may exist for each account number, legal organization, month, year, or other criteria. These totals are materialized views on journal entries in order to facilitate fast response times when aggregations are requested. Since column stores do not have to retrieve the whole tuple, but only the requested attribute when aggregating, we determined that can always be created on the fly. This makes aggregate maintenance obsolete and therefore reduces the number of aggregate updates to zero.

5.6.3.3 Status Updates

Status variables, for example, unpaid and paid, typically use a predefined set of values and thus create no problem when performing an in-line update since the cardinality of the variable does not change. If it is preferable to keep a record of status changes we can also use the insert-only approach in these cases. If the status variable has only two possible values, then a null value and a timestamp are the best option.

5.6.3.4 Value Updates

Besides aggregate and status updates, the value of a tuple (for example, the address of a customer) can be modified. Since the change of an attribute in an enterprise application has to be recorded (log of changes) in most cases, an insert-only approach seems to be the appropriate solution.

We analyzed customer data in different industries and the frequency of value updates in financial accounting applications covering a time span of about five years. The results are depicted in Fig. 5.16 and show that on average only 5% of tuples are actually changed.

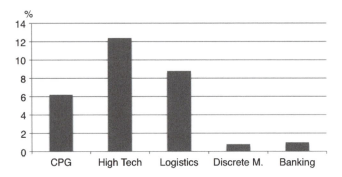

Fig. 5.16 Update frequency in financial accounting applications

5.6.3.5 Time Travel

Data can have multiple values at different points in time. For example, if the address of a customer changes, a change document has to be created to represent it within the database system. If the customer has some complaints about an invoice that was delivered to a former address, it must be possible to reconstruct the former address.

For financial applications, the requirement to retain all old (and updated) data is even enforced by law in some countries. To enable this, traditional applications use separate change documents that record all updates of specific database tuples. As a result, not only the actual master data table must be searched, but also the change document table has to be evaluated to retrieve an older version of the data.

With the insert-only approach on the other hand, data changes are recorded within the same database table, allowing more efficient queries of historical data. With the increased transparency of historical developments that insert-only offers, additional business questions can be answered. Figure 5.17 illustrates a simple time travel application with booked revenue in the past and revenue prognosis with 80 and 20% probability in the future, always based on the currently chosen point in time. The time slider (dotted line) enables movie-like browsing through history to see how prognosis and booked revenue differ at different points in time. Based on the insights of this development, better business decisions can be derived for the future.

5.6.4 Feasibility of the Insert-Only Approach

One of the disadvantages of the insert-only approach is the additional memory consumption. It is evident that the higher the update rate, the higher the additional memory consumption when using an insert-only approach, because for each row update, a new tuple has to be added to the database with the corresponding timestamps. Although efficient compression techniques can be applied within a column store

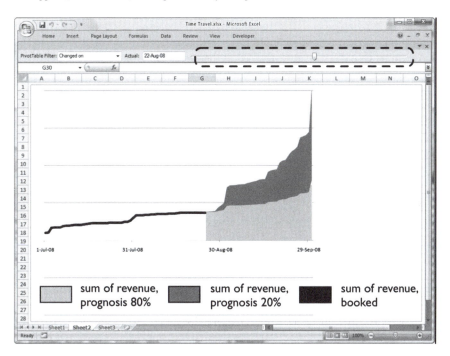

Fig. 5.17 Screenshot of a time travel application

(see Sect. 4.3), the memory consumption is still higher than updating the database tuples directly in-line.

In our customer study we observed an SAP financials application based on the SAP ERP suite. The system contains 28 million line items of the time period 2003–2007. Our observations are only based on the accounting document table (not including materialized aggregates and materialized views) and show the following behavior: of the 28 million line items in the accounting document table, there are 10.9 million journal entries that were reconciled through 1.4 million reconciliation entries. As a result, there are 12.3 million modified entries that—considering the original 28 million entries—would lead to a total number of 40 million entries in the accounting documents line items table when applying a traditional insert-only approach. This represents an additional memory consumption of about 40% [68]. However, the 10.9 million reconciliation updates, which are only status updates, can be handled with in-line updates and thus reduce the number of additional tuples to only 1.4 million, which is a 5% storage overhead.

Despite of potential higher memory consumption, there are many arguments that justify the feasibility of the insert-only approach. Besides the fact that typical financial applications are not update-intensive, by using insert-only and by not maintaining totals (materialized aggregates), we can even reduce these updates. Since there are less updates, there are less operations requiring locking, and the tables can be more easily distributed and partitioned horizontally across separate computing units (blades) with a shared nothing approach.

A number of changes can be applied to the financial system with the insert-only approach, which results in the fact that the number of major tables drops from more than 18 to 2 (not including change history and OLAP cubes): only the accounting documents—header and line items—are kept in tables. Combined with the fast aggregation algorithms executed on the fly, the insert-only approach replaces all indices, materialized views, and change history documents; and thus further reduces storage requirements, as outlined in the next section.

Another way to reduce the memory consumption is the concept of vertical partitioning for insert-only tables, as described in Sect. 4.4. The general idea is to isolate the parts of a relation that are changed often to reduce their negative impact. As a result, the increasing number of modified documents only has an impact on the partition in which they are stored. The unmodified partition is not changed. Depending on the different update frequencies for different attributes, we can show a clear improvement for the memory consumption [68]. The best case for using vertical partitioning is when the given relation has many attributes and only few of them are modified over time.

5.7 Enabling Analytics on Transactional Data

Unification of systems for transactional processing and analytical processing implies running both types of queries directly on transactional data. The problem that has led to the separation of the systems in the past was long-running and resource-intensive analytical queries influencing transaction throughput. With the advent of in-memory computing for enterprise applications, analytical queries can be executed on transactional data again [135].

One mission-critical step towards this goal is sufficient join and aggregation performance. Joins are required to combine data that has been spread over several tables due to the normalization of OLTP tables; aggregation is essential for analytical queries. We identified the following four techniques to be essential for fast join and aggregation performance:

- *Column-oriented storage*: because aggregations and joins operate along columns rather than along rows, storing relational tables column-wise avoids accessing data not required to answer aggregation and join queries.
- *In-memory storage*: with the amount of available main memory, it is possible to store the complete business data in main memory for analysis. Avoiding access to external memory promises significant performance gains.
- *Compression*: column-oriented storage is well suited to data compression, because all the values of a column have the same data type. This makes it easy to apply dictionary compression and other compression algorithms (see Sect. 4.3). Certain

join and aggregation implementations can even directly operate on the compressed data without decompression.

- *Partitioned storage*: large tables can be split horizontally and can be distributed across blades. Aggregates and joins can be computed in parallel on these distributed partitions.

This section gives an overview on how current hardware trends and the afore-mentioned advances in database technology can support the development of highly efficient join and aggregation algorithms for enterprise applications.

5.7.1 Aggregation on the Fly

Summarization is a natural approach to analyze large data sets. One way to summarize relational data is grouping and aggregation. Naturally, aggregation has become a core operation in data analysis tools. Examples can be found in OLAP, data mining, knowledge discovery, and scientific computing. In analytics applications of enterprise systems, aggregation typically operates on large amounts of data, and therefore the operation requires special optimization effort to cope with ever-growing data volumes. Allowing interactive ad-hoc analyses on the most recent enterprise data enables new types of applications (see Sect. 6.2). Such applications can run on mobile devices, where on-the-fly aggregation with fast response time is essential.

In this section, we introduce aggregation with the help of a simple example and give a classification of the various aggregation types. Then we introduce the basic aggregation algorithms. Based on these algorithms, we provide an overview of parallel variants.

5.7.1.1 Classifying Aggregation

When thinking about relational aggregation, most readers will probably have a SQL query similar to the following in mind:

Listing 5.4 (SQL Example)

```
SELECT Product, Country, SUM(NetSales)
FROM Sales
GROUP BY Product, Country
```

Logically, aggregations can be computed by a value-based decomposition of the input table into several partitions and by applying one or more aggregation functions to each partition. For example, the above query is evaluated by grouping tuples of input table Sales into partitions formed by the values of Product and Country: all tuples with the same values on these columns belong to the same partition, for each of which the SUM function is applied to the values of the NetSales column. Let us assume input data as shown in Table 5.4.

Table 5.5 Sales table

Product	Country	NetSales
Car	USA	$1,000,000
Car	Germany	$1,500,000
Boat	USA	$3,000,000
Car	Germany	$500,000
Car	USA	$100,000
Boat	Germany	$2,600,000
Car	USA	$1,300,000

Table 5.6 Aggregated values

Product	Country	NetSales
Car	USA	$2,400,000
Boat	USA	$3,000,000
Car	Germany	$2,000,000
Boat	Germany	$2,600,000

Then, the result for the above query would look like the data in Table 5.5.

As a convention, we will use the following terms throughout this section: columns, by which we can group, are called group columns, whereas columns containing values to aggregate are called measure columns. The number of rows in the result is called group cardinality in the following.

Aggregation cannot be expressed in Codd's initial relational algebra proposal [27]. However, with the introduction of SQL (containing aggregation as a central language construct), the operation also found its way into relational algebra. Besides the standard aggregate functions in SQL92, that is, SUM, MIN, MAX, AVG, and COUNT, several additional aggregation functions have been proposed to augment existing systems for statistic data analysis. Examples are MEDIAN, Standard Deviation, MaxN (the N maximum values), MinN, MostFrequentN (the N most frequent values), and the like.

We give an informal definition of the types of aggregation functions here:

- *Distributive*: per aggregate function and per group, a single field is required to keep the state during aggregate computation. Examples are SUM, MIN, MAX, and COUNT.
- *Algebraic*: per aggregate function and per group, a tuple of fixed size is required to keep the state during aggregate computation. Examples are AVG (to capture SUM and COUNT) or MinN/MaxN (to capture the N largest/smallest values).
- *Holistic*: the size to capture the state for an aggregation computation is only bound by the size of the group. An example for this class is MEDIAN. Per aggregate function and per group, $O(n)$ values have to be captured for groups of size n.

Further extensions have also been proposed for the grouping criteria: in our simple example above, grouping is defined on the equality of tuples on certain attributes. In general, we could apply any partitioning function, even those that allow the partitioning criteria to produce non-disjoint subsets. A common example often occurring in data warehousing is grouping by the nodes in a hierarchy, where a row is aggregated at different hierarchical levels.

While all these extensions are interesting, the common case supported in any relational DBMS are distributive and algebraic aggregate functions and disjoint partitioning criteria. We will focus on this standard case throughout the rest of this section.

5.7.1.2 Sequential Aggregation

Aggregation and join are related to the same problem, namely matching [60], that is, identifying all values in an input relation that belong together. To compute a join, we match values from at least two input relations. Join is classified as a binary matching problem, whereas aggregation is a unary matching problem. Another unary matching problem is computing all distinct values of a relation. For any kind of matching problem, three basic algorithms exist: the well-known nested-loops, sorting, and hashing.

To better understand parallel aggregation, we discuss sequential implementations of this algorithm and briefly summarize the three alternatives:

- *Nested-loops*: the nested-loops algorithm iterates over the input table to find all rows with the same value combination on the group columns. For all rows it finds, it calculates the aggregate function and writes the result into the output table. During processing, the algorithm keeps track of the value combinations on the group columns, for which it already computed the aggregates so that they are aggregated only once.
- *Sorting*: the sort-based algorithm sorts the input table by the values in the group columns. As a result, the rows that belong to the same group will be stored in consecutive order. The output relation can be computed by a single pass over the sorted result. Whenever a new group boundary is detected, the previously scanned values are aggregated and written to the output table. This algorithm performs best, if the table is physically sorted by values in the group column. Alternatively, it is possible to sort the table before the aggregation or to use a sorted index.
- *Hash-based*: the hash-based algorithm utilizes a hash table to build groups. For every row in the input relation, it creates a combined key from the values in the group columns. If the combined key is not present in the hash table, it is inserted together with the values to be aggregated. If the combined key is already present in the hash table, the aggregate is computed with the values already present.

The preferred solution in the general case is the hash-based approach, because the worst-case time complexity of this algorithm (depending on the number of rows in the input table) is linear. The sort-based algorithm has a worst-case complexity of $O(n \cdot log(n))$ and the nested-loops variant of $O(n^2)$, with n being the size of the relation. If the input is already sorted, then the sort-based algorithm should be applied, because the remaining aggregation part has a linear time complexity. Due to its quadratic behavior, the nested-loops variant should be avoided. In the following, we assume the general case and only discuss hash-based implementations for parallel aggregation.

5.7.1.3 Parallel Aggregation

To scale with the number of available processing resources, parallel aggregation algorithms have been proposed for shared-nothing architectures and for shared-memory architectures (we omit shared-disk systems). As explained in Sect. 4.2, modern in-memory database systems run on architectures with both characteristics: shared-memory within the blades and shared-nothing among the blades. We will introduce both types of algorithms. Let us look at shared-nothing aggregation algorithms first and let us assume that every computation node has its own part of the horizontally partitioned table to be aggregated. Three methods have been proposed to compute distributed aggregations [169]:

- *Merge-all/hierarchical merge*: every node computes the local aggregates based on its partition. All these local results are then sent to a central node, which receives them and sequentially merges them together to the final result. When many distinct group values appear, the intermediate size becomes large. In this case, due to sequential processing and high data load, the central merge node can become a bottleneck. An alternative is a hierarchy of merge nodes, where every node merges two or more intermediate results producing a single result, which is passed on to the parent merge node in the hierarchy. At all levels in this hierarchy, merges run in parallel, except for the final merge, which still assembles the result in a sequential fashion. In addition to these problems, intermediate result sizes might be skewed, resulting in unbalanced merge times and in slowing down the overall computation time.
- *Two-phase*: as in the merge-all approach, the first phase of the two-phase algorithm makes every node compute its local aggregated result. To distribute the merge load in the second phase, every node splits its local result by a globally known partitioning criterion. For example, we can define ranges on the group values according to which the local results are split (range-based partitioning). The disjoint partitions are sent to designated merge nodes, where each merge node is responsible for a particular partition. These merge nodes merge the partitions. Because the partitions are disjoint, the final result can be obtained by concatenating the results from the merge nodes. Merging is done in parallel. To keep the main-memory consumption of the algorithm low, the two phases can be interleaved: after some but not necessarily all intermediate results have been produced, partitioning and merging can already begin. Note that this algorithm cannot be used for holistic aggregation functions.
- *Redistribution*: this strategy first partitions and then aggregates. The partitioning criterion is defined on the group columns, such that every partition can be aggre-

gated independently and the results are disjoint. There are several possibilities for partitioning, for example, using value ranges or a hash function. The partitions are then sent to the aggregating nodes, and, finally, the results obtained at each node are concatenated.

The question is which alternative performs best. The answer depends on the result cardinality. If the cardinality is small, the first two approaches are superior, because the amount of data to be sent and merged is small compared to redistribution. If the number of groups is large, data transfer costs are high anyway, but the third alternative saves merge costs. Among the first two alternatives, the second one is advantageous, because it can avoid skewed merge times. Therefore, SanssouciDB implements alternatives two and three.

Distributed aggregation algorithms can run on shared-memory architectures without any change. In a distributed setting, algorithms for shared-nothing architectures need to move data from one node to another. On shared-memory machines, this is not necessary. To avoid unnecessary data movement, special parallel hash-based aggregation algorithms for shared-memory architectures have been proposed [26]. In the following, we will introduce the algorithm implemented in SanssouciDB.

On shared-memory architectures, multiple parallel threads can access the same data; that is, they can all access the input relation and they can share common data structures to store intermediate information or the result. In the following, we consider how the input data is accessed (see upper part of Fig. 5.18). Each thread (1) fetches a certain (rather small) partition of the input relation, (2) processes (aggregates) this partition, and (3) returns to step (1) until the complete input relation is processed.

We will call these type of threads aggregation threads in the following. The sketched range-based distribution approach avoids the problem of unbalanced computation costs among the aggregation threads (which is caused by skewed data distributions originating in the input partitions). Partitions with high computation costs are simply assigned randomly among the threads. Note that to implement the first step above, some kind of synchronization is required to ensure that each aggregation thread receives a disjoint partition of the input relation. This results in a trade-off between balancing computation costs (partition size) and synchronizing access to the input relation. Because assigning partitions to aggregation threads is a short-running operation (it simply computes ranges), aggregation threads will only block for a short period of time. During operation, each aggregation thread produces an intermediate result. The following three algorithms differ in the way they make use of shared data structures [26]:

- *Global hash table with synchronization*: all aggregation threads aggregate into a single shared hash table. To avoid inconsistencies, the access to the hash table has to be synchronized. Fine-grained locks (for example, per hash key) are required to enable concurrent insertions. Coarse-grained locks (one lock per hash table) would result in serialized insertions. Besides the overhead to maintain locks and the occasional lock wait time, global hash tables have another drawback: missing cache locality. Because a hash function computes an address for an input key, the likelihood of the according address reference to produce a cache miss is quite

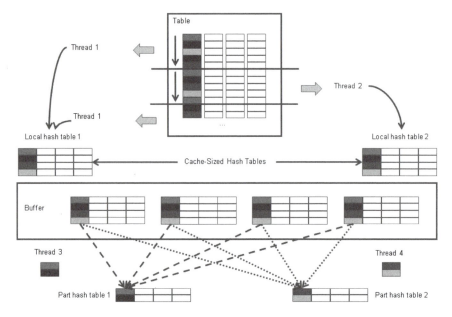

Fig. 5.18 Parallel aggregation with local hash tables

high (if the global hash table is sufficiently larger than the cache). As a solution, algorithms with cache-sized local hash tables have been proposed.

- *Local hash table with merge* (see Fig. 5.18): each thread has a private, cache-sized hash table, where it writes its aggregation results to. Since the hash table fits into the cache, misses are reduced. If the hash table is filled up to a certain threshold, for example 80%, the aggregation thread initializes a new hash table and keeps the old one(s) in a shared buffer. In the end, all buffered hash tables have to be merged. For the merge, we can apply the techniques discussed in the context of shared-nothing algorithms: merge-all/hierarchical and two-phase (with range partitioning and, optionally, interleaved merge). In contrast to the shared-nothing algorithms, mergers can now directly access the shared buffer. Figure 5.18 depicts the two-phase approach with range partitioning and interleaved merges. Thread 1 and Thread 2 are aggregation threads, writing into cache-sized hash tables. Thread 3 and Thread 4 are merger threads. The buffered tables are merged using range partitioning (each merger thread is responsible for a certain range). When the buffer reaches a certain size, the aggregation threads start to sleep and notify the merger threads. Every merger thread reads and aggregates a certain partition of all the hash tables in the buffer and aggregates them into a (yet again private) part hash table. The partitioning criteria can be defined on the keys of the local hash tables. For example all keys, whose binary representation starts with an 11 belong to the same range and are assigned to a certain merger thread.

- *Local/Global hash table with synchronization*: a combination of the two previous approaches leads to an algorithm that concurrently aggregates into local hash tables and occasionally merges the local hash tables into a single global hash table that is shared by all merger threads. Because the global hash table is not partitioned, merging has to be synchronized as in the first approach.

The second alternative (local hash table with merge) is implemented in SanssouciDB for two reasons: (1) memory consumption is restricted by the size of the buffer, and (2) when data is written, no synchronization is required, because every merger thread has a private hash table (in contrast to the other algorithms). When the input table is partitioned and distributed across different nodes, shared-memory aggregation can be combined with the shared-nothing techniques discussed above. The shared-memory algorithms can be used to obtain the node's local aggregation result. The optimal degree of parallelism, that is, the number of concurrently running threads for the aggregation algorithm depends on the query, the data, and on the current system workload. Based on the query and the data statistics, the plan generator makes a suggestion for the optimal parallelism degree for the aggregation operation. The system workload is monitored by the scheduler. To take the system workload into account, the scheduler can dynamically override the degree of parallelism suggested by the plan generator before the aggregation operation is scheduled (see also Sect. 5.10).

5.7.1.4 Aggregation in a Column Store

The way data is organized has a direct impact on aggregation performance. We already discussed the implications that partitioning across blades and the in-memory storage have and how they are reflected in our the aggregation algorithms. In the following, we want to highlight the implications of column storage on the shared-memory aggregation algorithm. As described so far, when processing a partition, an aggregation thread reads values from the group columns and from the measure columns row by row. Row-wise access is expensive in a column-oriented database system, because the data is scattered across different memory locations and is not nicely packed together, as in row stores (see Sect. 4.4). As a result, row-wise access would provoke many cache misses. To remedy this situation, we can use a special hash-table implementation called index hash map. The index hash map allows column-wise key insertion and column-wise aggregation. We will sketch the idea in the following. Assume we have a group column G containing four different keys (indicated by color) and a set of measure columns M_i to be aggregated, as depicted in Fig. 5.19a. First, column G is processed by the index hash map:

- Column G is inserted key by key. For each insertion, the index hash map returns an integer called index position. All keys with the same value obtain the same index position, for example, in Fig. 5.19b, the yellow value always obtains position 3.
- For every key, the corresponding index positions are kept in a mapping list (see Fig. 5.19c). Basically, the index position provides a mapping from the value position in column G to the group (represented by the index position) the value belongs to.

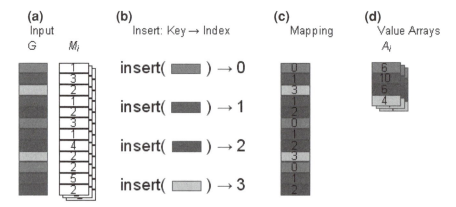

Fig. 5.19 The index hash map allows column-wise insertion. **a** Input; **b** insert: key → index; **c** mapping; **d** value arrays

- With this mapping, the aggregation can be calculated for each measure column M_i. The result is stored in a value array. The value array maps each group to a list of aggregated values, one list per measure column.

Note that the largest index position obtained from the index hash map indicates the number of distinct values D in column G. The underlying hash table implementation of the index hash map can be freely chosen. For example, we can apply bucket hashing, linear probing, or cuckoo hashing [111] to manage the key index-position pairs.

Having created a mapping vector for column G, the measure columns M_i can be processed in a column-wise manner to compute the aggregates: for every measure column M_i, a result array A_i of size D is created (see Fig. 5.19d). For the value at a certain position j in the measure column, we can obtain the index position k from the mapping vector at position j. The value is then aggregated at position k in array A_i.

The sketched algorithm has a small drawback when multiple group columns are requested. In this case, we have to feed a combined key into the hash table. Computing the combined key requires row-wise access to the group columns, which possibly produces cache misses in pure column stores. In a store with combined columns (see Sect. 4.4), where we can mix row-oriented and column-oriented storage, we ensure that the group columns that are often queried together are stored in a row-wise chunk to avoid these cache misses. However, such a physical optimization depends on the query workload and possibly needs expensive adjustments when the workload changes.

In a pure column store, we remedy the situation by preparing a special array, which holds a column-wise representation of the values from the group columns: during column insertion, gaps are left for the values of the following columns. From this array, we can then insert the combined group keys row by row. Note that this is only possible when the group columns have fixed-width values (for example, integer).

In SanssouciDB, dictionary encoding (see Sect. 4.3) always ensures this. One could ask, whether the overhead to implement column-wise aggregation is justified. The answer is that in a row store, complete rows have to be read by the processor to obtain the necessary columns for aggregation computation. Because main-memory I/O is the bottleneck, the overhead imposed by the column-oriented implementation is low compared to unnecessary data reads in the row-oriented alternative.

5.7.1.5 Aggregation in Data Warehouses

Good aggregation performance is one of the main objectives in data warehouses. One approach to achieve the necessary performance in ordinary relational data warehouses is to consolidate the data in cube data structure which is modeled as a star schema or snowflake schema (see Sect. 7.1). In a cube, expensive joins and data transformations are avoided and aggregation becomes a simple and cost-efficient scan over the central relational fact table. Due to the possible size of fact tables in real-world scenarios (a billion rows and more are not unusual), further optimizations, such as materialized aggregates, became necessary.

A materialized aggregate stores the answer to several aggregation queries at a certain level of granularity. Queries of the same or lower granularity can be answered by a lookup in the materialized aggregate. All other queries have to be evaluated on the base data. Materialized aggregates have several implications:

- Defining and maintaining materialized aggregates requires detailed knowledge of the application domain. Experts are required to choose the right dimensions and key figures for a materialized aggregate.
- Materialized aggregates require maintenance costs for data loading and synchronization with the base data. Synchronization takes place at certain intervals or on demand (lazy materialization). In the first case, the data represented by the aggregate does not reflect the most recent state. In the second case, response time performance decreases, because data has to be materialized during query evaluation.
- Materialized aggregates are not well suited for ad-hoc queries and interactive data exploration, because they are static data structures defined during design time and only cover a subset of all column combinations by which we can group.
- If a query does not match a materialized aggregate, the query is evaluated on the base data. In this case, a significant performance penalty has to be paid resulting to skewed query processing time.
- Materialized aggregates require additional storage space.
- Data warehouse system vendors have coding efforts to implement materialized aggregates and all required logic (for definition, maintenance, and query matching).

Obviously, materialized aggregates, the star schema, and the advanced storage model (compression, in-memory, column-oriented storage, partitioning) discussed above are orthogonal techniques: we could integrate them all in SanssouciDB. The question is however: do we still need a star schema and materialized aggregates if we can answer all aggregation queries from a column-oriented in-memory DBMS on the fly? With all the objections listed above, avoiding materialized aggregates (and possibly the need for the star schema) would dramatically reduce system complexity and TCO. As we stated in the introduction of this section, aggregation performance is a mission-critical factor for this goal. Past experiences have shown that in-memory technology really could make the difference. An example is the SAP Business Warehouse Accelerator (BWA), an extension to a classical data warehouse that can answer analytical queries from main memory without the need to predefine materialized aggregates. However, the BWA still requires the data to be stored in a star schema. With fast joins, we could answer analytical queries without the need to consolidate the data in a star schema. The queries would operate on the normalized tables and join the relevant information to be aggregated.

5.7.2 Analytical Queries without a Star Schema

Compared to the schema of most transactional data, which is commonly normalized in third normal form, the star schema is denormalized to reduce join processing costs. When executing analytical queries directly on the data in the source schema, it is crucial to have satisfactory join performance. In the following, we will take a look at how we can efficiently implement this information in a distributed and parallelized manner.

The relational join operator, commonly denoted by the symbol ⋈, combines tuples from two or more relations into a single relation. Logically, it computes the Cartesian product of all input relations and applies the join predicate to filter the tuples of the final result. In practice, joins are not evaluated this way, due to the performance penalty resulting from the Cartesian product. We will see in the following, how joins can be evaluated in an efficient way.

A join normally comes with a predicate, which can be any logical predicate defined on the attributes of a tuple. In most cases, the predicate checks the equality of some attribute values forming a so-called equijoin. Joins based on other comparison operators, for example, <, >, or != are called non-equijoins.

To illustrate the join operator, suppose we want to join the rows from tables SalesOrder R and SalesOrderItem S based on attribute OrderID. The data is shown in Tables 5.6 and 5.7.

In the following, we will call the columns based on which the join predicate is defined as join columns. The join predicate simply states that SalesOrder.OrderID shall be equal to the SalesOrderItem.OrderID. In SQL, this query would have the following form:

Table 5.7 Example table SalesOrder

OrderID	Date	SoldToParty
101	2010-10-01	FCA
102	2010-10-02	DSS
103	2010-10-02	AKM
104	2010-10-03	FCA
105	2010-10-04	HMM
106	2010-10-04	DSS

Table 5.8 Example table SalesOrderItem

ItemID	OrderID	ProductNumber
5004	101	510
5005	101	254
5006	102	301
5007	102	888
5008	103	890
5009	104	230
5010	105	120

Listing 5.5 (Join Example)

```
SELECT *
FROM SalesOrder, SalesOrderItem
WHERE SalesOrder.OrderID=SalesOrderItem.OrderID
```

The result is presented in Table 5.9.

In the Cartesian product, each row in R is combined with all rows in S. The predicate restricts the Cartesian product to the rows shown in Table 5.8. Note that the row with OrderID = 106 did not find any join partner in S. Therefore, no row was generated for this OrderID in the result. Of course, if we are not interested in certain columns, we can restrict the join result using a projection. The relational algebra symbol to express a projection is π_A, where A denotes the columns to retain in the projected result. Many join algorithms allow embedding projections to avoid materialization of unnecessary columns. A special join operation, which plays an important role in the implementation of distributed joins, is the semijoin. A semijoin is denoted by the symbol $\ltimes$ and restricts the returned columns to those that belong to one input relation (in this case, the left one), thereby removing duplicates according to relational theory. We can, for example, apply a semijoin to retrieve all rows from R, which have at least one corresponding sales order item in S (Table 5.9).

5.7.2.1 Sequential Join Processing

Because the join is a binary matching problem, we can apply the well-known nested-loops, sorting, and hashing strategies. We will first look at the basic sequential join

Table 5.9 Example result set

OrderID	Date	SoldToParty	ItemID	ProductNumber
101	2010-10-01	FCA	5004	510
101	2010-10-01	FCA	5005	254
102	2010-10-02	DSS	5006	301
102	2010-10-02	DSS	5007	888
103	2010-10-02	AKM	5008	890
104	2010-10-03	FCA	5009	230
105	2010-10-04	HMM	5010	120

implementations before we move on to distributed and parallel joins. We briefly sketch the following three alternatives:

- *Nested-loops*: the simplest way to implement a join is the nested-loops join. One relation (called the outer input) is scanned in the outer loop, while the other relation (called the inner input) is scanned in the inner loop. During scanning, both relations are checked for matches of the join predicate. If a match is found, the corresponding output tuple is constructed and written to the result. Because the inner input is scanned repeatedly, the performance of this algorithm is poor in most cases (its time complexity is $O(n \cdot m)$ with m and n being the sizes of the input relations). Nevertheless, if the result size is close to the size of the Cartesian product (that is, if the selectivity of the join predicate is low), nested-loops evaluation may provide sufficient performance. Some optimizations of the algorithm exist; for example, if the inner input is sorted on the join attributes or an appropriate database index is available, the inner loop can be replaced by binary search or an index lookup (index nested-loops joins).
- *Sort-merge*: the sort-merge join algorithm can be used for equijoins and some cases of non-equijoins. The algorithm consists of a first step, in which both relations are sorted on the join attributes, and a second step, in which the relations are scanned simultaneously to produce output tuples when matches are found. If there are multiple occurrences of a combination of joining attributes in both relations, one of the scans needs to return to the first occurrence of the value combination to construct all tuples of the output. The first step can be omitted for one or both relations if the relation has already been sorted in a previous step, or a database index enforcing a sort order on the joining attributes is available. The worst-case complexity of the algorithm is dominated by the sorting costs of the larger input relation (of size n); that is, $O(n \cdot log(n))$. Merging is linear with respect to the size of the input relations.
- *Hash-based*: hash joins can only be used for equijoins. The algorithm uses hashing to build a hash table by applying a hash function to the joining attributes of the smaller relation. The same hash function is used for the joining attributes of the larger relation to probe for a match while scanning over that relation. The hash join has a linear worst-case complexity depending on the size of its input.

The preferred solution in the general case depends on the query, the physical layout of the database (sorted columns, indices), the expected selectivity of the join predicate, and the amount of available main memory. Most database systems implement all three alternatives and apply cost-based selection of the best algorithm during the creation of the query evaluation plan. In the following, we describe join algorithms for distributed and shared-memory join processing.

5.7.2.2 Distributed Join Processing

SanssouciDB is a distributed database system. Relations can reside on different blades. To reduce the network traffic in distributed systems, a join operation can be expressed using a join of semijoins. Let us assume a join between tables R and S on the set of join columns A. For simplicity, we assume that the join columns have the same names in both relations. If tables R and S are stored on different nodes, the projection $\pi_A(R)$ of the join columns of R is calculated and sent to the node storing S where it is used to calculate $T = \pi_A(R) \bowtie S$; that is, the set of tuples of S that have a match in $R \bowtie S$. Projection $\pi_A(T)$ is sent back to the node storing R to calculate $U = R \bowtie \pi_A(T)$. The final join result is obtained by calculating $U \bowtie T$. This can be expressed in relational algebra as follows:

$$R \bowtie S = R \bowtie (R \bowtie S)$$
$$= R \bowtie (\pi_A(R) \bowtie S) = (R \ltimes (\pi_A(\pi_A(R) \bowtie S))) \bowtie (\pi_A(R) \bowtie S) \quad (5.3)$$

The last form computes two semijoins. For these two semijoins, it is only necessary to transport the values of the joining attributes from one node to the other. Values of attributes that are not part of the join condition only have to be transported for tuples that are in the final join result. SanssouciDB stores relations in a compressed store. To highlight the implications, let us look at our join example between tables SalesOrder and SalesOrderItem. To keep the presentation simple, we only show the two join columns in dictionary encoding (see Sect. 4.3) in Fig. 5.20. We assume that tables SalesOrder and SalesOrderItem are stored on different blades A and B. Let us assume the query shown in Listing 5.6:

Listing 5.6 (SQL Example)

```
SELECT *
FROM SalesOrder, SalesOrderItem
WHERE Date = '2010-10-02'
AND SalesOrder.OrderID = SalesOrderItem.OrderID
```

The distributed join can be computed in the following nine steps. The computation starts at blade A:

1. The ValueIDs that match the selection predicate are retrieved from the dictionary of the SalesOrder.Date column (just 01 in our example).

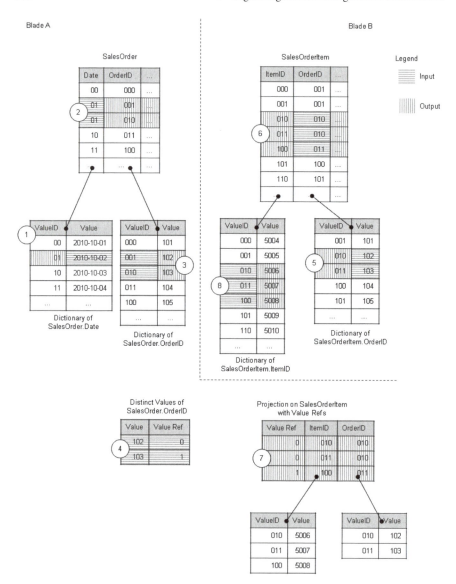

Fig. 5.20   A distributed join example

2. The list of all rows in the `SalesOrder.Date` column that appear in the `ValueID` list from the previous step is computed (rows 2 and 3 in our example).
3. Based on the list of rows from the previous step, the distinct values of the `SalesOrder.OrderID` join column are retrieved from its dictionary (`102` and `103` in our example).

4. The distinct values from the previous step are sent to blade B. Note, we add a virtual column to the list of distinct values called `Value Ref` for later use. This column is not physically present and is not sent to blade B.
5. At blade B, those value IDs from join column `SalesOrderItem.OrderID` are retrieved that correspond to the values sent from blade A in the previous step (`010` and `011` in our example).
6. The list of all rows in the `SalesOrderItem.OrderID` column that appear in the list of value IDs from the previous step is computed (rows 3, 4, and 5 in our example).
7. The dictionaries of all columns that participate in the final result are copied and reduced to the values that appear in the list of rows from the previous step.
8. A projection on the table at blade B is computed that contains the matching rows from Step 6. The reduced dictionaries are attached accordingly. Note that we do not need to attach a dictionary for the join column, because these values are already present at blade A. However, to be able to compute the final join at blade A, we add a column `Value Ref` that references the values in the distinct values list sent in Step 4. The entries are referenced by their position.
9. At blade A, the final join can be computed based on the projected table, the distinct values array, and the original table.

The advantages of a column store for the proposed algorithm are obvious: only the columns that contribute to the final result are accessed. Because joins operate on attributes, applying a filter on the join column (Step 2) and the calculation of distinct values (Step 3) only operate on the join column. In a row store, other columns would be read as well. Furthermore, filtering (Step 2) can be implemented very efficiently by exploiting SIMD processor extensions (see Sect. 4.2). The distinct-value calculation (Step 3) can be accomplished by the parallel aggregation algorithm introduced in Sect. 5.7.1, which achieves good cache locality. After this discussion of the shared-nothing algorithm for join processing, we want to introduce a shared-memory variant for blade-local parallel join computation.

5.7.2.3 Parallel Shared-Memory Joins

Just like aggregation, equijoins can be computed by hashing. In fact, to compute parallel shared-memory joins, we can easily re-use the infrastructure, that is, the index hash map introduced for parallel shared memory aggregation. The algorithm operates in two phases. In the first phase, the values of the smaller input relation's join attributes and their corresponding row numbers are inserted into the index hash map. This insertion is executed in parallel by concurrently running join threads. Again, each thread has a private cache-sized hash table that is placed into the buffer when full and that is occasionally merged into the part hash tables. The result consists of a set of part hash tables with key-value pairs. The pair's key corresponds to a value from a join column, and the pair's value corresponds to the list of row numbers indicating the position where the value occurs in the input relation. Again, the keys in the part hash tables are disjoint. In the second phase, the algorithm probes the join columns of the larger input relation against the part hash tables. Probing works as follows:

1. Private cache-sized hash tables for the larger input relation are created and populated by a number of concurrently running threads as described above. If the hash table is filled up to a certain threshold, for example 80%, it is placed in the buffer.
2. When the buffer is full, the join threads sleep and probing threads are notified. Each probing thread is responsible for a certain partition. Because the same partitioning criterion is used during the first phase of the algorithm, all join candidates are located in a certain part hash table. If the probing thread finds a certain value in a part hash table, it combines the rows and appends them to the output table.
3. The algorithm terminates, when the buffer is empty and the larger input relation has been consumed. The result is the joined table.

If a join is defined on more than one column, we have the same problems as during aggregation with multiple group columns: we need to create a combined key to feed into the hash table. As a solution, combined columns can be created (see Sect. 4.4). Alternatively, we prepare an intermediate array, into which the values from the group columns are copied column-wise, just like in the aggregation algorithm.

5.7.2.4 Advantages of In-Memory Column Stores for Join Computation

In disk-based database systems, the restricted amount of main memory made block-oriented data processing necessary. A block is the data transfer unit between disk and main memory. Therefore, the main problem with join operations in such systems is to organize block accesses in such a way that specific blocks are not accessed more frequently than necessary. The three most commonly used join algorithms have traditionally been implemented to optimize for block-based data access and to cope with insufficient main memory. In these implementations, intermediate results that do not fit into the main memory are written to disk.

In an in-memory database system like SanssouciDB, all data including intermediate results fit into main memory. Therefore, such a system does not have to cope with blocked data access anymore. Instead, algorithms that utilize multiple parallel threads and provide good CPU cache locality become important. Furthermore, such algorithms should only require short locks to synchronize data access. We have shown such an algorithm in this section.

As we have seen, column orientation also has a positive impact on join computation: just like aggregations, joins operate along columns. If two or more columns appear frequently as join columns in the query workload, SanssouciDB can store them together in a combined column simplifying hash-table insertion during the hash-based shared-memory join. Essential operations for distributed join processing, like computing the distinct values or applying a predicate, only need to access

Fig. 5.21 Example memory
layout for a row store

3523	Smith	Peter	3524	Carter
Arthur	3525	Brown	Karen	3526
Dolan	Jane	3527	Miller	Marc

the relevant column(s), furthermore, they can be parallelized as outlined above. In addition, columns are well suited for low-level SIMD operations and often result in a good cache locality. The combination of these aspects make in-memory column stores very well suited for computing aggregations and joins, two of the most commonly used database operations in enterprise applications.

5.8 Extending Data Layout Without Downtime

In all companies the schema and the underlying data layout of a database might have to be changed from time to time. This may happen when software is upgraded to a newer version, software is customized for specific needs of a company, or new reports must be created. Ideally, the data layout can be modified dynamically without any downtime of the database, which means that data is still accessible or updatable while the data layout is modified. In this section, we analyze data layout changes for row and column stores.

Also, in the case of multi-tenant databases, where multiple tenants are consolidated into the same database, each tenant might be interested in individually extending the standard data layout with custom extensions of the data model. This is difficult when tables are shared among different tenants. More details on this topic are presented in Sect. 8.3.1.

Research regarding dynamic schema modification has been undertaken particularly in the context of object-oriented databases [9, 10, 107]. In this section, we will focus on the impact of the data layout caused by schema changes for row-oriented and column-oriented database systems.

5.8.1 Reorganization in a Row Store

In row stores, operations changing the data layout, such as adding or removing an attribute from a table are expensive. In row-oriented databases, all attributes of a tuple are stored sequentially in the same block, where each block contains multiple rows. Figure 5.21 shows a simplified memory layout for a table containing customer information and the columns ID, Name and Given_Name. All columns are stored in the same memory block.

If no space is left at the end of the memory block, adding a new attribute to the table results in a reorganization of the whole internal storage. If a new attribute with

Fig. 5.22 Extended memory layout for a row store

3523	Smith	Peter	NV	3524
Carter	Arthur	NY	3525	Brown
Karen	CA	3526	Dolan	Jane
MI	3527	Miller	Marc	IL

Fig. 5.23 Example memory layout for a column store

3523	Smith	Peter	NV
3524	Carter	Arthur	NY
3525	Karen	Brown	CA
3526	Dolan	Jane	MI
3527	Miller	Marc	IL

the name State is added and there is not enough space at the end of each row, all following rows must be moved. Figure 5.22 shows the memory layout of the same block, after the attribute State has been added.

The same problem occurs when the size of an attribute is increased, for example when the maximum length of a string is increased from 50 to 100 characters. Reorganizing the storage for each block is a time-consuming process. Generally, it is not possible to read or change those fields while their memory layout is being reorganized. Row-oriented database systems often lock the access to the whole table, which leads to decreased service availability or delays. To be able to dynamically change the data layout, a common approach for row stores is to create a logical schema on top of the physical data layout [9]. This allows for changing the logical schema without modifying the physical representation of the database but also decreases the performance of the database because of the overhead that is introduced by accessing the metadata of the logical tables. Another approach is to introduce schema versioning for database systems [146].

5.8.2 On-The-Fly Addition in a Column Store

In column-oriented databases, each column is stored in a separate block. Figure 5.23 shows a simplified memory layout for the same table that stores the customer information.

Each column is stored independently from the other columns. New attributes, like State, can now be added very quickly because they will be created in a new memory area. In SanssouciDB, new columns will not be materialized until the first value is added. Locking for changing the data layout would only be required for a very short timeframe, during which the metadata storing information about existing columns and fields is changed. In a multi-tenant environment, it would also be easy to add custom columns for each tenant independently.

When an attribute is deleted, the whole memory area can simply be freed. Even when the size of an attribute is changed, only the dictionary for that special column must be reorganized, as shown in Sect. 4.3. As a result, changing the data layout of column-oriented databases results in less downtime and higher service availability than for row-oriented databases.

Particularly in multi-tenant systems, which will be discussed in Chap. 8, the need for adding additional columns arises frequently. In SanssouciDB, each tenant has its own metadata, which can be modified with custom extensions to the data model. Multi-tenant applications benefit from choosing a column database as the underlying storage.

When developing applications, schema changes occur often. If a row-based persistence is chosen for storing the data underneath the application, the reorganization of a database is expensive since all pages must be changed. Given its flexibility regarding schema changes, a column store is preferable for application development.

5.9 Business Resilience Through Advanced Logging Techniques

Enterprise applications are required to be both fault-tolerant and highly available. Fault tolerance in a database system refers to its ability to recover from a failure. It is achieved by executing a recovery protocol when restarting a failed database, thereby moving the failed database to its latest consistent state before the failure. This state must be derived using log data that survived the failure, that is, from logs residing on a non-volatile medium, for example Solid-State Drive (SSD). Although an in-memory database holds the bulk of data in volatile memory, it still has to use persistent storage frequently to provide a secure restart protocol and guarantee durability and consistency of data in case of failures.

To abstract from the vast number of possible hardware and software failures, recovery protocols generally assume halting failures. A halting failure describes any failure that leads to loss or corruption of data in main memory, be it the failure of an individual memory module or a power outage. For SanssouciDB, we assume an infrastructure service that detects halting failures, for example, by using periodical heartbeat messages, as well as a service that acts on detection of a failure, that is by restarting the database.

The processes of writing recovery information to a non-volatile medium during normal operation is called logging and consists of writing a continuous log file containing all changes made to the database. Since replaying a long log is expensive, checkpoints are created periodically that mark a point in the log up to which a consistent state resides on the non-volatile medium, allowing the recovery protocol to replay only log entries inserted after a checkpoint. Checkpointing is an effective mechanism to speed up recovery and well covered in the literature [17, 72]. Checkpointing strategies differ in how running transactions are affected while the checkpoint is taken, that is by being delayed. Logs must contain all necessary information to redo all changes during recovery since the last checkpoint. In case of a physical logging scheme, the physical address and the new value—the after image—is logged. Logical logging records the operation and its parameters, which results in larger log files than the ones generated with physical logging.

As in traditional disk-based database systems, the challenges for the in-memory databases are to (a) design logging mechanisms that reduce the frequency of I/O operations to non-volatile storage devices to a minimum in order to avoid delaying in-memory transaction processing and to (b) allow for fast recovery to keep database downtimes at a minimum. This is achieved by logging schemes that allow for parallel logging as well as parallel recovery. To increase throughput when writing logs to a non-volatile medium even further, a group commit scheme is employed. Here, the buffer manager waits for the commit of multiple transactions for persisting several commits at a time flushing the log buffer only once. Reload prioritization is a method to increase availability of the system by loading the most frequently accessed tables back into memory first, while tables with a low priority are loaded lazily.

Logging schemes for row-oriented data differ from logging of column-oriented data. SanssouciDB provides a persistence layer that is used by all logging schemes. This persistence layer implements a page-based buffer manager that uses write-ahead logging and shadow pages for uncommitted changes. The page-based buffer manager directly maps row-oriented data, which is organized in a page layout, and logging is done within the persistence layer. For column-oriented data, the persistence layer provides a log file interface, used by the column store for writing logs.

In the following we describe the logging and recovery mechanisms for column- and row-oriented data in the SanssouciDB.

5.9.1 Recovery in Column Stores

As described in Sect. 5.4, the column store consists of a read-optimized main store and a write-optimized differential buffer. Both structures are dictionary-encoded. The visibility of rows within the differential store is defined by the transaction manager, which maintains a list of changes done by transactions. Together, these three structures represent the consistent state of a table that must be recovered in case of failure.

The main part of a table is snapshot to a non-volatile medium when main store and differential buffer are merged. The merge process is a reorganization of the column store integrating main store and differential buffer as well as clearing the differential buffer afterwards as described in Sect. 5.4. In order to fall back to a consistent state during recovery, redo information for updates, inserts and deletes since the last merge are logged to non-volatile memory—the delta log—as well as the list of changes by uncommitted transactions—the commit log—are logged to non-volatile memory.

To recover a table in the column store, the main part is recovered using its latest snapshot, while the differential buffer is recovered by replaying the delta and commit log. While recovering the main store from a snapshot is fast, for example when using memory-mapped files on an SSD, replaying log information is the potential bottleneck for fast recovery. The time required for replaying the delta log increases linearly with the size of the log and depends on how well log processing can be done in parallel. Parallel processing of logs is possible if log entries can be replayed in arbitrary order, that is, when any order leads to the desired consistent state.

Like the main part of a table, the differential buffer consists of a vector holding value IDs and a dictionary containing a mapping of value IDs to the actual values. During normal operations, the dictionary builds up over time, that is, a new mapping entry is created in the dictionary each time a new unique value is inserted. Both differential buffer and dictionary must be rebuilt at recovery using the delta log.

In case of a logical logging scheme, recovery of the dictionary from the delta log requires replaying all log entries in the original sequential order for deriving the correct value IDs. Hence, to allow for parallel recovery, a logging scheme storing the insert position (row ID) and value ID is required. The logging scheme must also include logs for the dictionary, since new dictionary entries cannot be derived from logical logs anymore. Given these considerations, an insert log for a table of SanssouciDB's column store contains the transaction ID (TID), the affected attribute (attr), the row ID and the value ID of the inserted value. The TID is required for defining the visibility of the row for concurrent transactions based on the commit log. Such a logging scheme is sometimes also referred to as physiological logging [72], since the row ID can be interpreted as a physical address (the offset into the value ID vector) and the value ID as a logical log entry. As described in Sect. 5.4, the differential buffer is an insert-only structure and updates result in a new row. The row previously representing the tuple is marked invalid in the change list maintained by the transaction manager. To reduce the number of log entries for a row, only logs for attributes that are actually changed by an update operation are written. During recovery, the missing attribute values of a row can be derived from the row representing its logical predecessor record.

Although most updates are done using the insert-only approach described above, SanssouciDB also supports in-place updates for certain status fields. When performing an in-place update, the delta log contains multiple entries for the same row ID. In this case, only the row ID with the highest commit ID (taken from the commit log) is considered during recovery.

Figure 5.22 shows the logging scheme for two transactions TA1 and TA2. We disregard the commit log for the sake of clarity. Transaction TA1 starts before TA2 and inserts a new value a for attribute X in row 1, resulting in the log entry [1, X, 1, 1] and a log [X, a, 1] for the dictionary mapping 1 → a. TA2 writes values b and a causing a new dictionary entry b → 2.

To prevent dictionaries from becoming bottlenecks during transaction processing, dictionary entries are visible to other transactions before the writing transaction is actually committed. Thus, dictionary logs must be managed outside of the transactional context of a writing transaction to prevent removal of a dictionary entry in case of a rollback. For example, if TA1 is rolled back the dictionary entry 1 → a should not be removed because this value ID mapping is also used by TA2. In Fig. 5.24, this is depicted by not assigning TIDs to dictionary logs.

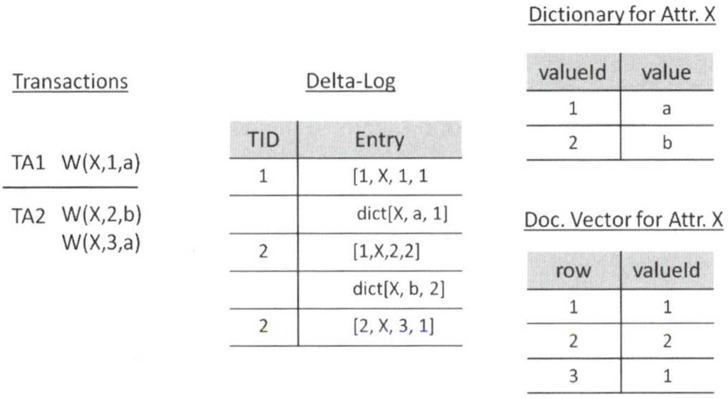

Dictionary for Attr. X

Transactions	Delta-Log		valueId	value
			1	a
	TID	Entry	2	b
TA1 W(X,1,a)	1	[1, X, 1, 1		

			Doc. Vector for Attr. X	
TA2 W(X,2,b)		dict[X, a, 1]		
W(X,3,a)	2	[1,X,2,2]	row	valueId
		dict[X, b, 2]	1	1
	2	[2, X, 3, 1]	2	2
			3	1

Fig. 5.24 Column store logging

The problem with logging dictionary changes outside the transactional context is that—in the event transactions are aborted—a dictionary might contain unused values. Unused entries are removed with the next merge of main store and differential buffer and will not find their way into the new main store. Since transaction aborts are assumed to be rare, we consider this to be a minor problem. To speed up recovery of the differential buffer further, the value ID vector as well as the dictionary can be snapshot from time to time, which allows for truncating the delta log.

5.9.2 Differential Logging for Row-Oriented Databases

Row-oriented data is organized in pages and segments (see Sect. 4.4). Pages are directly mapped onto pages persisted by the buffer manager. In contrast to the column store, changes made to segments of a page are logged using differential logging [103]. Since a change normally affects only parts of a page, a logging scheme is desirable, which considers only these changes, instead of logging the complete image of the page. Differential logging addresses this issue and additionally allows for parallel logging and recovery. Differential logs can be applied in an arbitrary order and only require to log one value instead of logging both, a before (undo) and after image (redo), as usually done when using physically logging [72]. A differential log only contains the XOR image of a fixed-length part of a memory page. The differential log is the result of the bit-wise XOR operation applied to the before image B and after image A of the page segment; we denote the differential log of A and B by $\Delta(A,B)$. The image might be written on record level or on field level granularity, depending on how many fields of a record were changed. Since the XOR operator is commutative, that is, (A XOR B = B XOR A), the order in which log entries are replayed does not matter. The XOR image tells for each bit whether it was changed or not. For example, if a specific bit is set in three XOR images, the bit will simply be toggled three times during replay, while the sequence is irrelevant. Additionally,

XOR is also an associative operator, that is, ((A XOR B) XOR C = A XOR (B XOR C)). This allows defining undo and redo operations solely based on the differential log Δ(A,B). For the after image A and the before image B, the undo operation is defined as A XOR Δ(A,B) while redo of an operation is given by B XOR Δ(A,B). Due to these properties of the XOR operation, logs can be applied in an arbitrary order, which allows for parallelizing both the logging and recovery process.

5.9.3 Providing High Availability

Availability is usually expressed as a percentage of uptime in a given year. For example, guaranteeing that a system is available 99.9% of the time (sometimes referred to as three nines) allows for at most 8.76 h of downtime per year and 99.99% availability (four nines) allows for at most 52.6 min of downtime per year. Such unavailability does generally not include downtime that is due to planned maintenance. High availability can be achieved in a number of ways. The two most common are as follows:

- *Active/Active*: multiple up-to-date copies exist, across which all requests are balanced. This is the approach adopted in SanssouciDB.
- *Active/Passive*: multiple up-to-date copies exist, but only one is active at a time. When the active copy goes down, the passive copy becomes the new master.

In order to make a process highly available, multiple copies of the same process (and its local data) must be created. The goal is to ensure that the data is still online in case one (or more) copies become unavailable. Besides, data replication provides several additional advantages, which are not related to availability but improve operating an in-memory database in general: it provides more opportunities for distributing work across the cluster by devising load balancing strategies that take replication into account; it can mask server outages (also with the aid of load balancing), both planned (upgrades) and unplanned (failures); and it can mask resource-intensive administrative operations, for example data migration (discussed in [156]) or the merge of deltas into columns. Here, we restrict the discussion to the issue of availability. Several mechanisms exist to keep multiple replicas up to date. They can be distinguished by how the writes are executed: synchronously or asynchronously. Synchronous writes ensure that once the data is written into the database, the data will be consistent in all parts or replicas of the written data set. It should be noted that network performance has drastically improved over the last year with the advent of technologies like Remote Direct Memory Access (RDMA) over InfiniBand, for example. Given that a write request over RDMA takes less than 3 μs within the same data center, the cost for synchronizing multiple replicas can be lowered significantly when using a high-end networking infrastructure. In case a link failure partitions the network in the data center or cloud, it might not be possible to write new data synchronously to all copies, unless the network is reconnected again. The only way to solve this would be to take the write functionality to the data offline and make it unavailable, or to postpone the update for the disconnected replicas, until the network is reconnected. In any case, the data would not be consistent in all copies, or it would not be available at all times. In contrast, asynchronous writes do not suffer from network partitions, since mechanisms for postponing writes and conflict resolution are a natural part of the asynchronous execution model. Even if a network partition occurs, there is no need to disable writes while waiting for the network to reconnect. There is a trade-off in keeping the system available and tolerant to network partitioning: the data of all replicas may not be consistent at all times, which simply is based on the fact that time is required to execute the write in an asynchronous fashion. Regardless of how many replicas of the same process (and data set) exist within a system, there has to be a way to recover failed instances. If the only instance holding the most current value being written to the database crashes before this value was propagated to other nodes in the system, the process must be able to recover this value and initiate propagation to other replicas, as we have described above.

5.10 The Importance of Optimal Scheduling for Mixed Workloads

In the following, we describe the basics of scheduling the execution of tasks amongst multiple available computing resources. In particular, we focus on scheduling strategies that are aware of the mixed workload, which is expected on SanssouciDB, as we outline our vision of an ideal scheduler for such scenarios.

5.10.1 Introduction to Scheduling

Assume we have a set of jobs (that can be broken into tasks) and several resources we can delegate to these tasks for processing. A job is finished when all its tasks are processed, but there are various ways to distribute the work amongst the available resources. Query processing in a parallel database system is an illustrative example for this problem: a query (job) consists of several operations (tasks) that can be executed by different processors or cores on various blades (resources). The question now is: how to assign tasks to the resources in an optimal way? This is what scheduling is about. Scheduling depends on a number of factors. Let us assume the trivial case: if the tasks have no inter-dependencies, each task utilizes its resource to the same extent (for example, it takes the same amount of time for completion), each job is executed exactly once at a previously known point in time, and there is at least one resource available per task when the task has to be executed, each task can simply be assigned to a separate resource. Unfortunately, the trivial case never occurs in real life.

In real life, the situation is more complicated. Let us look at this problem in the context of database query scheduling:

1. Tasks have inter-dependencies. For example, the probing phase in a hash join cannot run, until one input relation has been converted into a hash table (see Sect. 5.6). Also, operations in a query plan depend on each other.
2. Tasks may be triggered once or repeatedly by sporadic events, for example, a new query arrives at the database system.
3. Tasks have skewed resource-utilization characteristics, for example, applying a filter operation to different partitions of a horizontally split table may result in different result sizes and in different execution times.
4. A task may utilize different types of resources to a different extent. There are CPU-bound tasks like aggregation operations and memory-bandwidth bound tasks such as column or table scans in query processing.
5. Resources are restricted, for example, a restricted number of CPUs and a restricted memory bandwidth are available on the hardware the database system runs on.
6. Tasks may have affinities to processing resources or even hard requirements on them. For example, in a distributed database system, a filter task should be executed at the node where the data is stored.

To cope with these restrictions, a decision must be made about which task is executed when at which resource.[1] This means, a schedule has to be found. Usually, this is the task of a so-called scheduler, and the process of finding a schedule is known as scheduling. To find the best or at least a good schedule, scheduling decisions need to be optimized. Finding an optimal schedule is NP-hard in most cases, and to find the best solution would take too much time. Therefore, scheduling algorithms usually strive for some good solution to be found in an acceptable time span.

The problem space for scheduling is large and can be characterized by the task characteristics (Points 1–4), the machine environment (Points 5 and 6), and some kind of optimality criterion. What is optimal depends on the goal the system has to achieve. The most prominent goals are as follows: throughput (the number of queries processed in a certain time frame), latency (the time until a query is evaluated), and fairness. Unfortunately, these goals are sometimes mutually exclusive.

To give a simple example, consider two queries: query 1 consists of tasks A and B. Both tasks require 2 time units to run to completion. Query 2 consists of a single task C, which requires 50 units for completion. Let us assume that we can assign A, B, and C to only one computing resource (that is, the schedule is a sequential order of the three tasks). A fair schedule would be A, C, B, because the executions of the two queries are interleaved and each query is served to the same extent (Fig. 5.25). The average latency of this schedule is 53 time units, because Query 1 has to wait until task C of Query 2 finishes. A schedule optimized for latency would be A, B, C with an average latency of 27 time units. The third scheduling plan would be C, A, B, which is bad in average latency compared to the two other plans.

Scheduling concurrently running processes and threads is a classical operating-system task (since multi-tasking was introduced). Usually, operating systems strive

---

[1] There can be further restrictions, but we content ourselves with the ones relevant for the following discussion in the context of SanssouciDB.

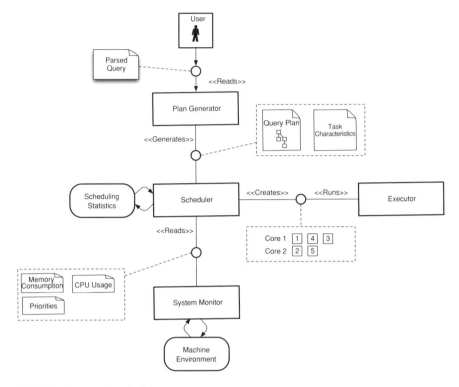

Fig. 5.25 Two sample schedules

for a good mix between throughput, latency, and fairness. Database systems can basically decide whether they want to rely on the operating system to schedule database processes and threads (OS-level scheduling), or whether they want to take over control (DBMS-level scheduling). In the first case, the database system simply creates as many processes and threads as necessary to process queries. The operating system then schedules these threads. In the second case, the database system starts and manages a fixed number of threads (usually as many as processor cores are available) and dynamically assigns tasks (for example, query processing operations) to threads. Note, in the first scenario, the number of threads created can easily be higher (oversubscription) or lower (undersubscription) than the number of threads that can be physically executed simultaneously given a certain number of CPUs (and cores.) Due to these problems, DBMS-level scheduling is preferable.

As explained above, a scheduler has to base scheduling decisions on the job characteristics J, the machine environment M, and the optimality criterion C. Modeling J and M to reflect a real-world system (for example, an in-memory database system running on high-end hardware) can be quite difficult. For example, the job characteristics of a simple filter task in SanssouciDB can depend on the selectivity of the filter, the data storage mode (row/column/delta), the size of the data (whether it fits nicely into the cache), the alignment of the data (for example, whether it provokes cache-line splits), and the storage location (for example, in a NUMA environment). Furthermore, modeling the machine environment on modern hardware (for example, the memory hierarchy) can also be complicated. Because scheduling decisions have to be computationally cheap with regards to the tasks they schedule, the scheduler cannot take arbitrarily complex models into account. Therefore, J and M usually abstract from many lower level details. In the following, we will discuss these general issues in the context of SanssouciDB.

5.10.2 Characteristics of a Mixed Workload

When using a main-memory database, optimal scheduling of tasks and optimal resource usage are mandatory to achieve the best possible throughput and latency. Typically, optimizing resource usage is a workload-dependent problem and gets even more complicated when two different types of workloads—transactional and analytical—are mixed.

With a focus on a transactional workload, one would ask if it is actually required to perform scheduling on such operations. The typical response time of a query in a transactional workload is usually less than a few milliseconds. If enough resources are available, the queue length (assuming that all incoming task queue globally at a single point) would be constant and the overall throughput would be sufficient.[2] However, if analytical queries are executed at the same time, the situation may change dramatically. The nature of analytical queries is that they touch a huge amount of data and perform more operations, so that both the CPU usage and memory consumption, for example, for intermediate results, is significantly higher compared to transactional queries.

Consider the following situation, where a given system is observed for 10 s. The system has 12 cores. For simplicity, we focus solely on execution time in this example. During those 10 s, two different queries are run repetitively: an analytical query A with an execution time of 2 s using all 12 cores and a transactional query T with an execution time of 0.1 s, running on only one core. If the queries A and T arrive sequentially in random order, we see two different problems: the analytical queries will acquire all system resources even though during the execution, new transactional queries arrive. If all queries now try to allocate the resources, more execution threads are created than resources are available. As a result, the operating system starts to schedule the execution (assuming OS-level scheduling), creating context switches, which slow down the overall performance.

Interestingly, when only observing the transactional queries, we would achieve 120 queries per second throughput without any scheduling. If this would suffice for our enterprise system, we would not have to implement scheduling at all. Most importantly, while the execution time for the analytical query might only be affected

---

[2] Given an even distribution of the workload and given an optimal placement of tables in memory.

minimally and the decrease in performance is more acceptable, for transactional queries, the situation is not acceptable, because the constant stream of new queries leads to resource contention.

In a second approach, a thread pool is used to share resources among all arriving queries and to perform active DBMS-level scheduling. Again, the analytical query allocates all threads while the transactional queries have to wait until the resources become free. Now the queue length increases and as a result, the system is rendered unable to handle the workload anymore.

This observation leads to the fact that resource management and query scheduling are important and heavily interconnected system components of mixed workload databases.

5.10.3 Scheduling Short and Long Running Tasks

The goal of this section is to outline important parameters for the implementation of a scheduling component and to describe strategies, which can be applied in a mixed workload environment. We follow the structure of Sect. 5.9.1 and first discuss the machine environment and task characteristics.

5.10.3.1 Machine Environment

System resources can be separated into time-shared resources like CPU and space-shared resources (for example, main memory). The most important resources that need to be covered during scheduling are:

- *CPU*: how many compute resources, that is, cores are available?
- *Memory bandwidth*: what is the available memory bandwidth that is shared by all compute units during query execution?
- *Main memory*: how much main memory is available to be scheduled?
- *Other*: what kind of other operations (for example, network communication) have to be considered during query execution, if the database system, for example, is distributed?

SanssouciDB's target hardware described in Chap. 3 gives us some insight. Essentially, we assume a blade cluster consisting of high-end blades, each equipped with 64 cores and 2 TB of main memory. The remaining parameters of the machine environment can be determined using an automatic calibrator tool that experimentally retrieves machine characteristics. The resulting machine model can be used as an input to the scheduler. In the following discussion, we leave out scheduling of system resources in a distributed database setup. We see that parts of the common single-site concepts can be applied equally well for a distributed scenario. In a distributed setting, it is crucial to carefully design the scheduling component because it can easily become a bottleneck and single point of failure.

5.10.3.2 Task Granularity

As stated above, scheduling is based on the execution of a job that can be broken down into tasks. In a database system, such tasks can have different levels of granularity. The more fine-grained the schedule, the closer to optimal the resource consumption will be. On the other hand, fine-grained tasks require more scheduling decisions resulting in a higher scheduling overhead. A good trade-off is required. The naïve approach to setting task granularity is to use the query as a scheduling unit. While this might be suitable for short-running transactional query plans, for long-running and even distributed analytical queries, too much scheduling-relevant information is hidden inside the query. During query execution, the original query is decomposed into a query plan that is executed to generate the expected result. Since the plan operators are the smallest execution entity, they provide a good foundation as an atomic scheduling unit, for example, as a task. Note that the task dependencies (see Point 1 in Sect. 5.9.1) are naturally given by the query's operator tree.

5.10.3.3 Task Characteristics

During plan creation, when the overall query is subdivided into a query plan, statistical information about the nature of the plan operations is collected to optimize the query execution. To optimize the scheduling decision, such parameters are also of interest for the scheduling algorithm.

In an in-memory database system, plan operators have the following properties that can be used for scheduling:

- Operator parallelism describes the best possible parallelization degree for the plan operator. This number determines how many concurrent threads access a given memory location.
- Memory consumption determines how much memory is scanned during the plan operation. This parameter can be used during the scheduling decision making process to estimate how much memory bandwidth is consumed.
- CPU boundedness determines, if the operator implementation is limited by the number of compute resources (for example, hash probing).
- Memory boundedness determines, if the operator implementation is limited by the available memory bandwidth (for example, scanning).
- Core affinity determines on which processor a specific task should be executed. This information is especially useful in NUMA architectures.
- Cache sensitivity—determines if the operator is sensitive to CPU caches. This parameter can be used to hint to the scheduler that, if two cache sensitive operations are run at the same time, the execution might result in worse execution performance, since both operators depend on the cache size (for example, hash probe, hash build).

The above parameters can now be used as input for the scheduling algorithm to best partition the set of operators among the shared resources.

5.10.3.4 Scheduling Vision

Based on the system observation and the observation of the enterprise workload, the scheduler has to cope with the two different workloads. The basic optimization goal is to guarantee the flawless execution of all OLTP transactions since they provide the basis for all operations of the enterprise applications. This means that the application has to define a certain threshold for transactional queries that defines a minimal acceptable state. The remaining available resources can now be scheduled for the incoming analytical queries. The flexibility of the scheduler allows increasing or decreasing the degree of parallelism for the queries based on the system load. One possibility to achieve a separation between those two resource areas could be to have two different dynamic thread pools. One thread pool represents the resources required for the OLTP transactions and the second thread-pool represents the resources available for analytical queries.

The scheduler now observes the system's performance statistically and is able to move resources from one pool to another, but always making sure that the minimum throughput for transactional queries does never fall below the given threshold. Using this approach the scheduler can dynamically adjust to the usage profile of the above applications. A typical example for such an observation could be that during the last weekend of the month almost no transactional queries are executed while many different analytical queries are run. Here, the system observes the month-end closing and can now adjust the resource distribution for this period so that more resources are available for analytical queries.

However, a key property of the scheduling component is to make a near-optimal scheduling decision in the shortest possible time frame. Existing work already provides good approaches to scheduling in a mixed workload system. The more parameters are added as inputs to the scheduling algorithm, the bigger becomes the solution space, resulting in an increased time to find the optimal scheduling decision.

If a good scheduling algorithm is applied in the system that allows the addition of scheduling parameters, one could consider adding a task priority parameter (as found in operating system scheduling) to increase the probability that a process is executed during a certain time frame (known as nice-value of a process). For enterprise computing there are several applications of such priorities:

- Priority of requests from mobile applications versus workplace computers.
- Priority of requests issued from a certain user group versus normal queries.

The priority could then be used to move a request up in the input queue or even allocate more resources for the request, for example, if the scheduler would assign only four threads for an analytical query, with special priorities the degree of parallelism could be increased to a higher level.

5.11 Conclusion

In SanssouciDB, data organization is not only focused on providing good access performance for generic database operations. SanssouciDB also addresses the specific requirements of enterprise applications.

One of these requirements is that data-intensive operations can be executed close to the data. SanssouciDB allows application programmers to express algorithms in stored procedures. Compared to implementing an algorithm in the application layer, stored procedures can reduce network traffic for data transport. To speed up certain common queries, SanssouciDB also allows the creation of inverted indices.

Another requirement specific to enterprise applications is data aging. Because companies usually store data for a period of ten years but only actively using the most recent 20%, it is makes sense to partition the data into an active part and a passive part. The passive part is read-only and can be kept on a slower storage medium, such as flash disk. The application can specify the aging behavior because every application knows when some particular data item can be flagged as passive. The transition of a tuple from the active part into the passive part is then managed transparently by SanssouciDB.

Enterprise applications manage business objects. These business objects are hierarchically structured and are stored in a set of relations in SanssouciDB. To speed up the retrieval of a business object, our database system creates an object data guide? an index structure that allows the efficient construction of the business object from the relational store.

In the compressed column store of SanssouciDB, updates and inserts would provoke frequent recompression operations. To avoid these expensive operations, SanssouciDB has a write-optimized differential store into which updates are written. From time to time, the differential store has to be merged with the main store to optimize for read-intensive operations. In this chapter, we have seen how this merge process is implemented and how a column-wise merge strategy can ensure a low main-memory footprint.

The insert-only approach allows SanssouciDB to keep all committed data changes for processing. This enables business applications to query the database state at an arbitrary point in time (time travel). Besides supporting business applications, the insert-only approach integrates well with multi-version concurrency control: because, due to insert-only, all versions are kept in the database.

A necessary prerequisite for enabling analytics on transactional data is good join and aggregation performance. To scale with the number of cores and the number of blades, SanssouciDB provides algorithms for distributed and shared-memory join and aggregation calculation. The shared-memory variants are based on hash tables, which are aligned to the cache size to minimize cache misses. The distributed algorithms are optimized for minimal data transfer across blades.

Customizing enterprise applications often requires the adding of new columns to an existing database. In contrast to a row store, the SanssouciDB column store allows adding or removing columns without data reorganization at the record level. Therefore, these operations can run without downtime.

To enable fault tolerance and high availability of enterprise applications, SanssouciDB implements a parallel logging and recovery scheme and allows active replicas to take over in case of failures to increase availability. The logging scheme is a physiological logging approach that does not enforce a specific ordering of the log entries on restart. This allows logs to be written to multiple log files in parallel and leverages the increased throughput of multiple I/O channels during normal operations as well as during recovery.

A database system like SanssouciDB has to run many tasks concurrently. In such a setting, relying on OS-level scheduling can be disadvantageous; because running more threads than there are physical cores available can lead to expensive context switches. Therefore, SanssouciDB has to run exactly one thread per core and has to apply DBMS-level scheduling. The SanssouciDB scheduler has to be aware of the mixed analytical and transactional workload. The idea is to manage separate thread pools, one for analytical queries and one for transactional queries. Based on system monitoring, the scheduler can dynamically re-adjust the number of threads in each pool.

Part III
How In-Memory Changes the Game

Chapter 6
Application Development

Abstract In Part II we described the ideas behind our new database architecture and explained the technical details it is founded on. In this chapter we describe how we can best take advantage of the technology from an application development perspective. We first discuss how changing the way enterprise applications are written to take advantage of the SanssouciDB architecture, can dramatically improve performance. We then go on to give some examples of prototype re-implementations of existing applications, to show that the use of in-memory technology can reduce the complexity of the applications, as well as improve their performance.

Taking advantage of the performance benefits offered by an IMDB means redesigning applications to take advantage of the specific strengths of in-memory technology. The main way to achieve this is to push work that is currently done in the application layer, down to the database. This not only allows developers to take advantage of special operations offered by the DBMS but also reduces the amount of data that must be transferred between layers. This can lead to substantial performance improvements and can open up new application areas.

6.1 Optimizing Application Development for SanssouciDB

With the characteristics presented in Part II, SanssouciDB provides the foundation for the development of new kinds of enterprise applications. But the database is only one part of the overall application architecture. To fully leverage the possible performance improvements, the other layers of the architecture need to be developed with the characteristics of the underlying database system in mind.

In this section, we give an overview of the aspects that need to be considered when creating applications on top of a DBMS like SanssouciDB. Using a small example, we demonstrate that a careful choice of where to execute different parts of application logic is critical to performance. We conclude the section with general guidelines for application development when using an in-memory database system.

H. Plattner and A. Zeier, *In-Memory Data Management*,
DOI: 10.1007/978-3-642-29575-1_6, © Springer-Verlag Berlin Heidelberg 2012

6.1.1 An In-Memory Application Programming Model

In-memory DBMS technology disrupts the way enterprise applications have to be developed. In this section we discuss the fundamental principles of a new programming model, which enables fast and effective use of the advantages in-memory DBMSs, such as SanssouciDB, have to offer. In general, the programming model introduced aims to define and enable optimal usage of the underlying in-memory concepts with regard to performance on the one hand, and usability with a focus on programmability and developers' productivity on the other hand. The latter is often neglected when it comes to discussing database architectures, but has to be taken into consideration in order to leverage the fundamentally different concepts presented in this book to their full potential [4]. In the following, we discuss the architecture of the system with regard to the interfaces between applications and the DBMS from a programmability point of view. We then take a look at how the development of applications changes in general and what best practices and tools are either available or necessary in order to guarantee optimal usage of the database when developing a new application, and to support developers in their daily activities. Finally, we examine the necessary capabilities every application programmer has to have in order to effectively work with an in-memory DBMS such as SanssouciDB.

6.1.1.1 System Architecture from the Viewpoint of Application Development

Transferring large result sets from the database to the application can account for a major fraction of the overall cost of a query in terms of latency and therefore it is a goal of the system architecture to minimize this. The chosen approach as described in Sect. 5.1.3 tries to keep those functions of enterprise applications that consist of sequences of queries whose results are input to the subsequent query close to the database, in order to reduce cost for transferring the data back and forth between the application and database layers. This is possible by allowing for application logic to be implemented close to or even within the database via easy extensibility of database functionality in the form of stored procedures written in procedurally extended SQL, called SQL script in SanssouciDB. Such procedures are usually defined at design time.

In addition to stored procedures, which are pre-compiled and meant for multiple execution, possibly by different applications and thus can be seen as a static extension of the database's functionality, complex business logic may require dynamic generation of procedures to be executed within the DBMS itself. Recent research in the field of database programming environments (see [71]) proposes supporting the formulation of queries in procedural languages, which can be compiled by the DBMS at runtime in order to avoid performance loss through interpretation. A first step towards the support of dynamically generated procedures in SanssouciDB can be achieved by automatically generating SQL script, which is then deployed in the form of stored procedures. Additionally, low-level access to the database's data structures by application code is possible via explicit data structures—this of course requires explicit knowledge of the database's internal data structures.

6.1.1.2 Application Development

Most application developers feel comfortable with an object-oriented programming paradigm. An important goal is to increase developer productivity so this paradigm should be followed as far as possible in any new programming model. From a database point of view, the most important part of enterprise application development is the interaction with the database in the form of SQL queries as described in Sect. 5.1. To cope with this mismatch of object-oriented programming on the one hand and SQL as the lingua franca for the communication with the database on the other, an Object Relational Mapper (ORM) should be used. For operations on a single business object, SQL code can be generated by the ORM. Set operations that require manual optimization should be written in SQL by the application developer. It is important that this ORM is optimized to work with SanssouciDB. This includes the creation and use of appropriate database views. They represent the logical interface between application developer and database layer. Using these views saves the developer from being burdened with creating complex SQL queries accessing multiple tables.

In our system the ORM converts requests into pure SQL while other DBMSs also offer DB access via procedural programming languages (such as Java, ABAP Open SQL). The fundamental difference between a high-level declarative programming language, like SQL, and a procedural programming language is how the DBMS has to handle them. In a declarative programming language, the developer describes what the result set should look like and is not concerned with the control flow within the DBMS needed to materialize it. A major aspect therefore is, that after parsing the SQL statement the DBMS itself has to plan the execution of the query and can therefore optimize the flow while the application remains agnostic to the organization of data and data structures within the DBMS. A single SQL statement may contain multiple operators and its execution may require the calculation of intermediate results, which are only needed within the database and do not have to be transferred to the application. However, formulating complex application logic in a declarative style necessitates the creation of a single statement that describes those and only those records to be returned, which is often difficult to develop, maintain, and debug. It should also be noted, that a developer has to have explicit knowledge of the schema of the data within the DBMS to write declarative code.

In contrast, in a procedural programming language, the developer gives a step-by-step routine that extracts the data from the tables and translates them to the required format. Intermediate results can be pruned and combined with each other step-by-step, possibly resulting in more readable code, which is therefore easier to debug and maintain. The major drawback of procedural programming is that it is very difficult for the database to optimize automatically, as this would require a semantic understanding of the application's algorithm. Also, creating intermediate results and working on those may introduce significant overhead depending on where the application logic is processed, as transmitting intermediate results back and forth to the database layer requires bandwidth, memory, and computing power and increases query latency (see Sect. 6.1.1.1).

6.1.1.3 View Layer Concept

While all data stored in SanssouciDB can be used for operational and analytical workloads, multiple logical subsets may exist. The subsets provide an application specific view on the data while avoiding data redundancy. Views store no data, but only transformation descriptions, for example, computations, possibly taking data from several data sources, which might be other views or actual database tables. Any application on the presentation layer can access data from the virtually unified data store using views. For analytical workloads, virtual cubes similar to views exist, providing the same interface on real cubes but using the original data instead of a materialized copy.

We call this model the view layer concept. Using this concept, the developer does not need to have a complete overview of all existing data during the development process. Views allow the developer to focus on a specific subset of data and decreases the complexity of database queries for the developer and the user. Complex join and filter conditions are already applied within the view itself. Because these views are not materialized, the developer can modify a view and evaluate the resulting implications on his application in real time instead of having to wait until the data is materialized.

6.1.1.4 Guiding Principles and Tools for Developing Applications Backed by an In-Memory DBMS

When analyzing the necessity for tools for application development we see two major areas. First, there is a need for supporting development of applications by enabling the programmer to easily access the database to formulate and test queries (see Sect. 6.1.5). Second, we think that it is essential to provide a set of tools that focuses on performance optimization, i.e., minimal overall query runtime, at the time of application development. Such tools are not yet available off the shelf and integrated with an IDE, however, we see their functionality covered by the following aspects.

- *Visualization of query runtime during application development* While formulating an SQL query against a complex database schema, its runtime is not obvious to the developer. The query's general cost complexity, as well as the impact of specific parameters has to be made available so that these can be optimized on the fly and in an iterative process, as minimal changes to an SQL statement, for example, the set of attributes that needs to be materialized may have a substantial impact on its execution and the size of the result set. Additionally, the number of times a specific query is executed, which is dependent from the control flow of the application and its interactivity with the user, have be taken into consideration when estimating a query's runtime during the execution of the application.

- *Usage of real data in the development process* Monitoring the overall runtime of queries requires executing them against a dataset that accurately represents the complexity, in terms of the schema and size, of the data that will be used in the productive version of the application.
- *Establishing minimality of results sets* Analysis of application code shows that result sets transferred to an application often are not minimal with regard to the computation they are needed for. A common example is the number of attributes transferred, e.g., "SELECT * FROM table" versus "SELECT attribute FROM table" when only a single attribute is needed by the application's logic. First, this poses unnecessary overhead on the database itself for materializing tuples, which is more costly in column-oriented architectures. Second, the unused data transferred between the database and the application as part of the result set leads to higher latency and memory consumption. Determining the usage of queried attributes within an application is straightforward up to the point where it is dependent on the control flow what the minimal result set would have been. Furthermore, we see examples where results are used to perform calculations that could have been carried out in the database, for example, in order to check for the existence of a specific value within a column. Some of these cases can be detected within application code via pattern recognition and hence can be automatically resolved. Other examples require semantic understanding of application logic and are thus complex to resolve.

6.1.1.5 Requirements of the Application Developer

Developing applications on top of an in-memory DBMS such as SanssouciDB requires that application developers have the responsibility to constantly monitor and optimize the performance of the interaction with the database.

As described in Chap. 5, it is an essential part of the concept of SanssouciDB to access its data via declarative programming languages. First of all, this requires general knowledge of SQL and possibly of SQL script. Results from qualitative interviews with developers of enterprise applications switching from conventional disk-based row-store databases to SanssouciDB revealed that there is a need for education in order to leverage the full potential of SanssouciDB's functionality. Second, formulating declarative queries requires explicit knowledge about the data schema within the database. This includes naming conventions, for tables and attributes, as well as foreign-key relations and the existence of indices and views.

Furthermore, it is essential that the use of available tools with the functionalities as discussed in Sect. 6.1.1.4 are fully integrated into the development processes at all times, i.e., during implementation and integration, during testing, and also for optimizing deployed systems if the underlying data changes over time.

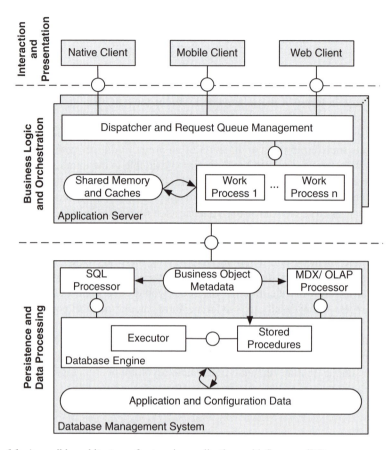

Fig. 6.1 A possible architecture of enterprise applications with SanssouciDB

6.1.2 Application Architecture

As initially presented in Sect. 2.3 and displayed in a more detailed manner in Fig. 6.1, enterprise applications conceptually comprise three layers. The interaction and presentation layer is responsible for providing a compelling user interface. This includes the creation of simple views that are aware of the current working context of the user and provide all required information in a clear and understandable fashion.

The business logic and orchestration layer serves as a mediator between the user interface and the persistence layer. It receives the user inputs and performs the appropriate actions. This can either be the direct execution of data operations or the delegation of calls to the database. To avoid database calls for frequently requested, yet rarely changed data, this layer may maintain a set of caches.

Data persistence and processing is handled in the bottom layer of the application stack. This tier provides interfaces for requesting the stored data by means of declarative query languages, as well as performing operations to prepare data for further processing in the upper layers.

6.1.3 Moving Business Logic into the Database

Each of the above layers can be found in all enterprise applications, but multiple factors determine their extent and where their functionality is executed. Section. 2.3 already outlined that, for example, parts of the business logic can be moved closer to the client or to the database depending on the performance characteristics of the respective subsystem.

Predominantly analytical programs will rely on a very thin orchestration layer that only translates the user inputs into corresponding SQL or Multidimensional Expression (MDX, see Sect. 7.1.5 for more details) queries. These queries are handled by the respective processing engine of the DBMS and are then forwarded to the query executor. In such a scenario, it is likely that most of the business logic is integrated into the client application and application servers only act as proxies for the database, if they are needed, at all.

In the following, we compare two different implementations of an enterprise application that identifies all due invoices per customer and aggregates their amount. The goal is to illustrate the impact of moving business logic into the database. The first solution implements business logic entirely in the application layer. It depends on given object structures and encodes the algorithms in terms of the used programming language. The pseudocode for this algorithm is presented in Listing 6.1.

Listing 6.1 (Example for the Implementation of a Simple Report)

```
for customer in allCustomers() do
 for invoice in customer.unpaidInvoices() do
  if invoice.dueDate < Date.today()
    dueInvoiceVolume[customer.id] +=
      invoice.totalAmount
  end
 end
end
```

The second approach, which is presented in Listing 6.2, uses a single SQL query to retrieve the same result set.

Listing 6.2 (SQL Query for a Simple Report)

```
SELECT invoices.customerId,
    SUM(invoices.totalAmount) AS dueInvoiceVolume
FROM invoices
```

```
WHERE invoices.isPaid IS FALSE AND
   invoices.dueDate < CURDATE()
GROUP BY invoices.customerId
```

For the first approach an object instance for each open invoice needs to be created to determine, if it is due. During each iteration of the inner loop, a query is executed in the database to retrieve all attributes of the customer invoice. This implementation has two disadvantages: for each created object, all attributes are loaded, although only a single attribute is actually needed. This leads to unnecessary data transfer as the algorithm re-implements database logic that can be more efficiently expressed in SQL.

The second implementation moves the algorithm into the database layer, to execute the business logic as close as possible to the data. Thus, only the required attributes are accessed and the number of data transfers between application and database is minimized.

When evaluating both implementations against small, artificial test data sets, this overhead looks negligible. However, once the system is in use and gets populated with realistic amounts of data, that is, thousands of customers with multiple open invoices per customer, using the first method is will be much slower. It is crucial that implementations are not only tested with small sets of test data, but against data sets and workloads that are representative for the ones expected in customer systems.

Another approach for moving application logic into the database are stored procedures, which were introduced in Sect. 5.1. Their usage for complex operations provides the following benefits:

• Reduced data transfer between the application and the database server.
• Pre-compilation of queries increases the performance for repeated execution.
• Reuse of business logic.

Stored procedures are usually written in proprietary programming languages or in proprietary SQL dialects such as PL/SQL (Procedural Language/SQL). However, many vendors also incorporate the ability to use standard languages (C++, Python or Java, for example) for the implementation.

In the context of the aforementioned scenario, creating a stored procedure to encode the required functionality is straightforward (see Listing 6.3). Since no parameters are needed for this application, it suffices to use the respective commands for defining a stored procedure and encapsulate the SQL code that was introduced in Listing 6.2. If applications want to use this functionality, they no longer have to define the complete SQL statement, but only call the stored procedure. Please note, that this example is intentionally kept very simple. By using invocation parameters and other features of the programming language used to define stored procedures, much more complicated functionality can be encoded and executed on the database level.

Listing 6.3 (Stored Procedure Definition and Invocation for a Simple Report)

```
// Definition
CREATE PROCEDURE dueInvoiceVolumePerCustomer()
BEGIN
 SELECT invoices.customerId,
     SUM(invoices.totalAmount) AS dueInvoiceVolume
 FROM invoices
WHERE invoices.isPaid IS FALSE AND
   invoices.dueDate < CURDATE()
GROUP BY invoices.customerId;
END;


// Invocation
CALL dueInvoiceVolumePerCustomer();
```

A drawback of current DBMSs is that, within stored procedures, only the database schema is known. Relations and restrictions that are encoded in a business object meta-model of the application are typically not accessible here. In contrast to such standard stored procedure implementations, SanssouciDB provides access to a meta-model of the application and its objects during query execution.

6.1.4 Best Practices

In the following, we summarize the discussions of this section by outlining the most important rules of thumb when working with enterprise applications in general, and with SanssouciDB in particular. We do not claim the completeness of this list of recommendations, but from our experience we believe that these are the most important ones when working with such systems.

- *The right place for data processing* during application development, developers have to decide where operations on data should be executed. The more data is processed during a single operation, the closer it should be executed to the database. Set processing operations like aggregations should be executed in the database while single record operations should be part of the application layer.
- *Only load what you need* for application developers it is often tempting to load more data than is actually required, because this allows easier adoption to unforeseen use cases. This can lead to severe performance penalties, if those use cases never occur. We advise to make sure that developers are aware which and how much data is loaded from the database and that these decisions are justified.
- *No unnecessary generalization* this recommendation goes hand in hand with the previous paradigm about data loading. One important goal for application developers is to build software that can easily be adapted to newer versions and allows incremental addition of new features. However, this requirement can lead to unnecessary generalization. This happens when developers try to predict possible future

use cases of the software and accordingly add, for example, interfaces and abstractions, which may load more data than currently required. This combination leads to a decrease in application performance and we advise to carefully add new features and to discuss added features in the development group.

- *Develop applications with real data* to identify possible bottlenecks of the application architecture, we generally advise to include tests with customer data as early as possible in the development cycle. Only with the help of this real data, it is possible to identify patterns that may have a negative impact on the application performance.
- *Work in multi-disciplinary teams* we believe that only joint, multidisciplinary efforts of user interface designers, application programmers, database specialists, and domain experts will lead to the creation of new, innovative applications. Each of them is able to optimize one aspect of a possible solution, but only if they jointly try to solve problems, the others will benefit from their knowledge. If their work is strictly separated, user interface designers, for example, might not be aware of the capabilities of the underlying DBMS and artificially limit their designs.

6.1.5 Graphical Creation of Views

With the support for ad-hoc query model creation and the increasing amount of data that is processed by an in-memory database, the requirements for the support of data model creation and its documentation increase as well. In this section, we discuss the needs for graphical database modeling and view creation within an in-memory database during the development and lifetime of the database.

6.1.5.1 Collecting Meta Data

New applications use two different kinds of data: data already that already exists in the database and data that is currently not present.

Data is organized in views, containing meta data about structure, complexity of joins, descriptions of the produced results and access right definitions. The information about which views are used by what applications is also stored in the database. Using this information, the developer can find the data already present in the system.

Data, that is not presently in the system needs to be defined. New tables will need to be annotated with the expected data growth rate over time. This data is used for performance predictions of selections, joins and aggregations. This helps the database to optimize the data layout and to trigger quality alerts if one or many of those development assumptions do not match the behavior of the real life system.

Fig. 6.2 Interactive real-time performance prediction during database view creation

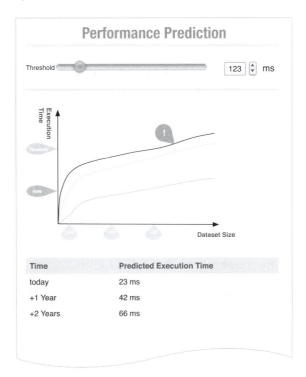

6.1.5.2 Intelligent View Creation

While a view defines a structured subset of the data available in database tables, creating a view for an application can be complicated. Views hide the complexity of joining multiple tables and enable the developer to access the data with less complex queries. Thus, the complexity moves from the query into the view. The developer can use graphical tools to create views. Those tools should be able to automatically determine the attributes used for joining tables, using naming conventions and dictionary comparisons. The modeling tool can automatically compare the dictionaries of the table's attributes to determine which attributes represent the best match for a join.

6.1.5.3 Interactive Performance Prediction

The modeling tool should support the developer by predicting the performance of the created view, as well as executing queries defined by the view creator. An example of such a performance prediction is shown in Fig. 6.2. While the developer can set a specific performance threshold for his own view, the modeler predicts the performance for the query execution. To get a more detailed understanding of system behavior this prediction is split into three parts: data access time, execution time of other views used in the view definition, and view execution time. While the data access time is mainly based on the size of the database, the execution time of other views and the currently defined view is based on the view definitions themselves.

Based on the query execution speeds provided by the tool, the actual performance of a created view can be determined by real-time execution of the view. Furthermore, the system will automatically provide information about future execution times of the view based on historically recorded system behavior, like data growth rate. It will also inform the developer at which point in time the set performance threshold will not be met anymore based on the current assumptions. This features becomes possible because the performance of in-memory databases can be predicted by less complex models than those used to predict the behavior of disk-based database systems [157, 162].

6.1.5.4 Reducing the Number of Layers

The view layer concepts enables the developer to define logical views based on other views. While this reduces the amount of time required to define new views, it will increase the number of views that have to be evaluated during query processing, decreasing the precision of predicted performance measures. To circumvent these issues, physical views exist next to the logical views. These are built from the logical view definition, while pre-compiling their execution paths.

6.1.5.5 Conclusion

The usage of logical views enables SanssouciDB to run transactional and analytical workloads on the same dataset without the need for data replication. We believe that using graphical tools to create, modify and monitor views increases the productivity of application developers and results in a steeper training curve.

6.2 Innovative Enterprise Applications

In the following section, we introduce new applications made possible by the use of the in-memory database concepts we described in Part I and Part II. The applications discussed in this section fall into one of the two categories. We present completely new applications that were not possible without in-memory technology due to timing constraints and we share how existing applications were improved through the use of in-memory technology, for example, in terms of scalability.

Our in-memory technology enables the design of current and forthcoming applications. New types of applications can be created now because limiting factors, for example, slow disk access, data exchange between the database and the application, and iterative database queries have been removed. Applications based on in-memory technology process large amounts of data or inputs events while providing sub-second response time.

We now discuss a number of applications, focusing on their characteristics in the following three layers as mentioned in Sect. 6.1:

- Interaction and presentation.
- Business logic and orchestration.
- Persistence and data processing layer.

6.2.1 New Analytical Applications

In the following, enhancements for analytical applications are presented. In contrast to nowadays applications, which exchange large amounts of data between database and business logic and orchestration layer, these applications perform the required set processing in the database. The benefit is that data is processed where it is located without being transferred through the entire application stack for processing.

6.2.1.1 Smart Meter Readings without Aggregates

We consider a Smart Grid as an energy network that can react to actions of different users, for example, generators, consumers, and those doing both. In a Smart Grid each household has a smart meter that records electricity consumption in intervals that can be read-out remotely [158].

Today's enterprise world is characterized by more and more data from various devices, processes, and services. Any of these individual data sources is a kind of sensor generating data in a constant flow. Aggregating all these sensor data can help to systematically measure, detect, or analyze data trends, which can then be used to derive new information.

For energy providers, the operation of Smart Grids comes with an initial financial investment for smart meters and the associated infrastructure components, for example, event aggregators and systems for meter data unification and synchronization. Smart meters are digital metering equipment that can be read without the need for on-site metering personnel. These devices can support services for switching off households when moving out, by performing required operations autonomously. The use of smart meters reduces variable costs for metering personnel and technical experts. Energy providers are able to offer new flexibility to their consumers. Instead of receiving an invoice once a year or on a monthly basis, consumers are able to track their energy consumptions in real time online. By using in-memory technology, getting billing details in real time becomes feasible. Each time the consumer logs in, corresponding readings are aggregated as needed.

We analyzed the viability of using in-memory technology in the context of Smart Grids in various projects, to handle the huge amounts of sensor data in new enterprise applications. For example, we used a master energy profile, describing the mean energy consumption of all households in Germany, to simulate specific meter readings for millions of consumers. We generated individual annual consumption profiles for each household and stored them on disk. Each file can be identified by an eight-digit unique profile number and consumes approximately 1.6 MB on disk. Our findings with SanssouciDB show that by applying in-memory technology to persist this data, storage consumption can be compressed by a ratio of about 8:1. In our test scenario with 500,000 customers, annual energy data consumed approximately 750 GB on disk. As a result of using the compression techniques described in Sect. 4.3, we were able to store compressed data in the main memory of a single machine where it consumes only 95 GB, a fraction of the initial data volume. Metering data for a specific customer can be identified with the help of a smart meter identifier and a certain timestamp. New metering data is appended to the list of readings without the need to update existing entries. This outcome verifies that it is possible to store metering data entirely in main memory with the help of SanssouciDB. The technology foundation for smart grids is built today to handle the huge amount of metering data of every household, consumer, and industry.

Our findings showed that any kind of on-demand aggregation in our prototype for a specific consumer was performed in less than one second. In other words, whether we are querying the energy consumption for an entire year, the last 24 h, or comparing two separate days with each other, it is possible to do this in an interactive manner. We were able to show, that SanssouciDB is a feasible technology to handle smart grid event data while providing analytical functionality with sub-second response time. This ability enables completely new applications, where consumers individually select the data to show and how to combine it; for example, users could analyze their personal energy usage in the energy provider's online portal. The latter is part of the interaction and presentation layer and only aggregated results are returned to it via the business logic and orchestration layer. The task performed by the latter is responsible for transforming data from the database to an adequate user interface format. For a single consumer, we recorded the following measurements: the annual energy consumption—an aggregation of 35,040 readings—is calculated within 317 ms, and the consumption for a single day—an aggregation of 8,640 readings—is available within 331 ms. Adding new values can be performed without locking and does not block any operations that are performed in parallel. Detailed results are available in literature [158].

Metering data is mainly required to perform billing runs. Aggregating the energy consumption of a time interval to create an invoice is performed on a set of metering data for a certain consumer. Set operations are invoked on data directly within the database and only the aggregated result is returned to the application.

We consider the given enterprise application for Smart Grids as an analytical example since metering data is aggregated for billing runs, allowing us to detect top consumers and to compare consumptions of multiple days. The performance of this application originates from the fact that the necessary set processing is performed directly in the persistence and data processing layer. It is only a single aggregated value returned to the application rather than sets of data.

Our further research activities show that real-time processing of sensor data can also be applied to further industry areas, for example, Supply Chain Management (SCM) with the help of Radio Frequency Identification (RFID). The use of RFID technology for tracking and tracing of goods is characterized by timing constraint similarly to the aforementioned since hundreds or thousand of reading events are created when goods are sent, received or during goods movements. We were able to verify that real-time authentication of tags and anti-counterfeiting of goods in the pharmaceutical supply chain benefit from the real-time analysis capabilities provided by SanssouciDB [159, 160]. From our perspective, anti-counterfeiting for a global RFID-aided supply chain is not possible without in-memory technology when a meaningful response time is required. For example, when items are unloaded from a truck and passed through reader gates, database latencies during verification of individual goods would result in delays of goods receipt processing.

6.2.1.2 Analytical Purchase Decisions

Our field studies in a franchise retail market revealed that procurement could be a time-consuming operation. If order decisions are made automatically based on statistical analyses, seasonal products might be ordered too late or in the wrong quantity. In contrast to fully automated purchasing, which is carried out once a day by server systems, semi-automated purchasing is applied to costly or seasonal products. For the latter, automatically generated purchasing proposals based on current forecasts must be verified, which is due to external weather conditions, before purchase orders are triggered. Our interviews with a German home- improvement store showed that product suppliers use their own personnel to check the availability of their products and need to make purchasing decisions on current quantities and delivery dates in each market by their own personnel. This decision involves two aspects: controlling product quantities in stock, which is supported by retail market management software; and a good knowledge of the market, to estimate the quantity of products that will be sold before the next purchase is scheduled. A manual check is then carried out, which includes the retrieving point of sales figures to estimate the validity of the purchasing proposal. This process was not supported by the specific enterprise application used by our interview partner.

External purchasing personnel in retail stores use small personal digital assistants to retrieve sales details of a certain product for the last month, and to acquire purchasing details. Due to security concerns and bounded hardware resources, queries are limited to a small subset of pre-aggregated data retrieved from the retailer's back-end systems. A new application based on in-memory technology can bridge the gap between fully automated and semi-automated purchasing. The challenge here is to provide the latest data without latency, to evaluate purchasing proposals and to offer the flexibility provided by BI systems. Our application prototype based on SanssouciDB combines current sales details with flexible functionality in the business logic and orchestration layer generally only provided by a data warehouse, for example, individual aggregations on current data. The use of SanssouciDB extends the view on data. On the one hand, pre-aggregated totals are eliminated and summed data is queried only when it is needed. As a side effect, fine-grained details about the relevant product instances can be accessed at the same speed, without additional latency, since they have already been traversed and are thus available for instant aggregation. On the other hand, external retail personnel can decide which period and product to analyze to support their retail decision. This operation is now performed on a product instance level rather than on an aggregate level for a class of products. The in-memory application performs its operations on the subset of relevant data for a product supplier. Identifying the relevant subset of data is performed directly within the database using stored procedures and does not involve interaction with the application. In other words, once the subset of relevant data is identified in the persistence and data processing layer, the application works only on this small portion.

This prototype focuses on supporting external retail personnel by allowing them to optimize their purchasing decisions. Seasonal products suffer from underprovisioning when their demand is high. For example, snow shovels are only available in limited supply when snow starts falling. This prototype aims to support better purchase decisions in retail stores, which are supplied by external purchase personal. In this prototype, the performance improvement results from the fact that only the selected subset of data that is accessed by a certain supplier's personnel in contrast to the total sales data for all items in the retail store need to be traversed.

6.2.2 Operational Processing to Simplify Daily Business

In the following section we share our prototype experiences in the area of operational applications. The execution of these applications results in operational processing enhanced for end users, for example, due to improved response time or processing of individual product instances rather than product categories.

6.2.2.1 Dunning Runs in Less Than a Second

We consider the dunning business process to be one of the most cost-intensive business operations. Performing dunning runs in current enterprise landscapes has an impact on the OLTP response time because of the large number of invoices and customers accounts that need to be scanned. To reduce the impact on the response time of the operational system, dunning runs are performed during the night or over weekend. Our field research at a German mobile phone operator revealed that dunning runs in current enterprise applications can only be performed at daily or weekly intervals for all customers to balance incoming payments and open invoices. It would be helpful to balance open invoices in an interactive manner for individual customers just when they are calling. Dunned customers get blocked from cellular services until the problem can be resolved.

The dunning process is mainly a scan of a long table containing millions of entries which have a certain dunning level for each entry. In response to a certain dunning level, predefined actions are triggered, for example, service blacklisting. We transferred a selected subset of dunning relevant data of the mobile phone operator and stored it completely in main memory. For our prototype we adapted dunning algorithms to use this new storage engine. Our tests show that in-memory technology is able to improve execution time of the dunning run by more than a factor of 1,200. For the given subset of real customer data, we experienced an execution time of less than one second compared to initial 20 min it had taken with the old system. This outcome shows that in-memory technology is capable of improving the response time of existing applications by orders of magnitude.

In-memory technology can be used to implement completely new business applications. Since we achieved response times of less than one second, the former dunning batch run is transformed into an interactive application. For example, a check can be performed instantaneously for individual customers when they are calling the operator's telephone hotline to validate her/his money transfers.

Managers can query the top ten overdue invoices and their amount on their personal cellular phone, notebook, or any mobile device with an Internet connection at any time. This information can be used to directly send a short message, an e-mail, or a letter to these customers.

We consider the optimized dunning run as an operational business example due to the fact that customers with overdue invoices receive a dunning notification. The performance improvement resides in the persistence and data processing layer since set processing is moved from the application closer to the data stored in the database as a stored procedure, which is executed within the persistence and data processing layer.

6.2.2.2 Availability Checks Reveal the Real Popularity of Products

Planning is a key business process in SCM [94], especially for manufacturing enterprises, as it is essential to satisfy their customers' demands. It is the starting point for cost reduction and optimizing stock keeping, for example, reducing the number of expensive products in stock. Improving existing applications used for planning has a direct impact on business operations.

We decided to move the primary storage of the planning data in our application to main memory while the execution of planning algorithms was distributed over several blades with multiple CPU cores per system.

Besides planning algorithms, we focused on ATP, which provides checking mechanisms to obtain a feasible due date for a customer order based on the items contained in the order. Modern SCM systems store relevant data tuples as pre-aggregated totals [113], which consumes additional processing time when updating or inserting new data. Redundant data requires hardware resources, which cannot be used for

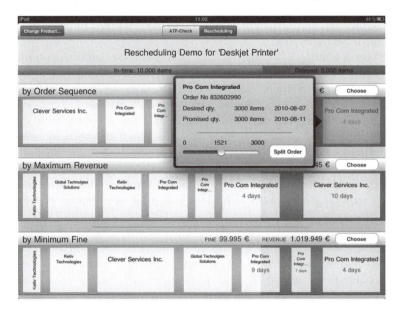

Fig. 6.3 Apple iPad prototype: interactive availability overview

processing purposes. Existing ATP check implementations suffer from performance drawbacks when including fine-grained product attributes for all products. Keeping response time at an adequate level for interactive work, product details are simplified to keep data amount at a manageable level.

Our prototype, as shown in Fig. 6.3, was designed with the following aspects in mind. Pre-aggregated totals were completely omitted by storing all individual data tuples in SanssouciDB. For a concrete ATP check this up-to-date list is scanned. We implemented a lock-free ATP check algorithm. Eliminating locks comes with the advantage that data can be processed concurrently, which works especially well in hot-spot scenarios when multiple users perform ATP checks concurrently for different products.

For the prototype implementation we used anonymized real customer data from a Fortune 500 company with 64 million line items per year and 300 ATP checks per minute. Analysis of the customer data showed that more than 80 % of all ATP checks touch no more than 30 % of all products. A response time of 0.7 s per check was achieved using serial checking. In contrast, we were able to execute ten concurrently running checks with our adapted reservation algorithm in a response time of less than 0.02 s per check.

This approach eliminates the need for aggregate tables, which reduces the total storage demands and the number of inserts and updates required to keep the totals up to date. The insert load on the SanssouciDB is reduced and this spare capacity can be used by queries and by instant aggregations. Since all object instances are traversed for aggregation, fine-grained checks on any product attribute are possible. Current

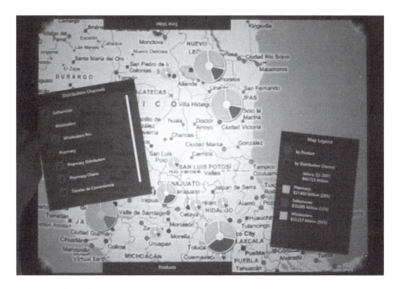

Fig. 6.4 Microsoft surface prototype

ATP implementations are kept lean by focusing on a minimum of product attributes. For example, printers of a certain model with a German and an English cable set are grouped together or both cable sets are added to improve performance. With the help of SanssouciDB, printers including different cable sets can be considered as individual products during ATP checks.

6.2.3 Information at Your Fingertips with Innovative User-Interfaces

In the following, we present synergies that become possible when combining innovative user interfaces with traditional applications powered by SanssouciDB. The use of modern design approaches introduces a new level of flexibility. From the user's perspective, predefined reports or application views limit personal working behavior. We have learned that using in-memory technology eliminates the need for fixed, predefined application designs and introduces a tremendous level of flexibility for the user.

6.2.3.1 Analyzing Sales Data in Real Time Using Microsoft Surface

A typical visualization pattern when discussing enterprise applications is the grid layout displaying data. In terms of innovative application design, business users of enterprise applications desire more interactive and flexible user interfaces.

The prototype in Fig. 6.4 shows geo-visualization of sales data combining two worlds: a modern user interface provided by a Microsoft Surface table and real-time analysis functionality processing 64 million sales figures stored in SanssouciDB. The application shows a sales region as a map, which can be interactively moved and modified by a single or multiple users working collaboratively together at the table. The user's interaction with the Microsoft Surface table results in analytical queries for sales data. Rather than specifying fixed SQL statements or executing a predefined report, as it is the case with standard enterprise applications, required queries are automatically composed by our application prototype based on the user's interaction at the table. The corresponding queries are derived and processed for the sales data stored in SanssouciDB. The results of all queries are returned in less than one second to enable interactive operations with the table. This application is only possible due to the use of in-memory technology and has special requirements for response times. As the user navigates through the map, zooms in or zooms out, queries are automatically triggered. The analytical application benefits from aggregating relevant sets in SanssouciDB and returning only summed data to the calling application.

6.2.3.2 Improved Navigational Behavior in Enterprise Applications

We analyzed the navigational behavior of business users and found that different users tend to navigate through enterprise applications in different ways to achieve the same goal. We developed a prototype application called C'estBON that offers flexible navigation through enterprise applications without the need to follow predefined paths. Semantically related business entities are identified from various data sources, for example, ERP systems, Microsoft Office applications, Customer Relationship Management (CRM) systems, or mail communications. After identifying semantic connections, it is possible to shortcut the navigation between application levels; for example, one can directly switch from an invoice to the related order or one of the ordered products. This ad-hoc navigation is based on the assumption that the application offers a way of navigating through a hierarchical structure, which is often the case for current enterprise applications. The C'estBON prototype is shown as a screenshot in Fig. 6.5 bypassing the predefined navigation of SAP Business One.

The implementation of C'estBON becomes possible by making use of a common data model describing various kinds of business entities, for example, invoices, orders, or business customers in a consistent way. Our approach combines the navigational aspect known from hierarchical data and applies it to transactional data. This example shows two aspects. On the one hand, existing transactional data has to be traversed in real time while opening a view of an enterprise application. This becomes possible by making use of in-memory technology. On the other hand, it builds on a data model defined by underlying business processes. Data needs to be processed the moment it is received from its source.

An application like C'estBON would be difficult to implement without in-memory technology because of the interactive time constraints and the interaction between

Fig. 6.5 C'estBON prototype: SAP business one with C'estBON

individual instances rather than object classes. Traditionally, all transactional data would need to be replicated to a dedicated OLAP system and an optimization of the data schema for the possible navigation paths would be necessary. With in-memory technology, however, this replication is no longer necessary and instead of navigating through potentially outdated data, all queries are performed on the actual OLTP system and still satisfy the given interactive time constraints.

C'estBON supports indirect navigation paths between instances of various classes, for example, a concrete sales order, its order positions, and the material details for a certain product. Its user interface is used to retrieve the user's mouse events while the business logic and orchestration layer constructs relevant queries.

6.2.3.3 Mobile In-Memory Applications

Mobility does change the way we communicate; even further, it changes the way we live and work. In fact, mobility is already transforming enterprises and the way they are run. Embracing the consumer trend, enterprises are increasingly unwiring themselves. We see a new kind of workforce evolving, the *mobile information worker*. Mobility is fast becoming a regular way of life in business, with mobile devices being the preferred interaction point to access and send information to any device, anytime, anywhere. Businesses realize that a mobilized workforce is more productive, since employees can now do their job on any device, at any time, and from any location. This capability not only incrementally improves the productivity of existing processes, but also provides a vehicle to revolutionize existing processes, resulting in dramatic productivity improvements.

Despite its potential, mobility also poses challenges to the enterprise. While technical concerns, such as security, data latency, and device management are particularly prevalent but not related to the usage of in-memory database management, other concerns such as usability play a central role. People expect the same ease of use they get from their consumer mobile app experience. In an unwired enterprise, "mobile is the new desktop"—it connects the boardroom to the shop floor and on to the consumer across the entire supply chain, it empowers people and the companies that employ them, and changes our culture and the way we work and interact with our customers. The in-memory technology supports a very important paradigm of mobility which is *interactivity*. Without an adequate response time, mobile users would not accept applications and use them extensively. Despite the latency of mobile networks, response can still be within one second and can thus allow interactive handling of massive data sets.

An example of such a mobile application is a mobile prototype that has been developed to support the sales force of Hilti, one of the world's leading manufacturers of power tools for the professional construction and building maintenance markets, with real-time information in customer dialogs [187]. The developed iPad application enables mobile sales representatives, potentially being at the customer site, to inquire about product availability in nearby inventories and receive cross-selling recommendations. Having this information available in real-time enables the sales force to tell customers instantly when an order can be shipped and provide relevant product recommendations based on historic sales data. This way, organizations can leverage their operational data to support their sales force in the field and thus, achieve a competitive advantage over rival companies.The usage of the in-memory database made it possible to achieve acceptable response times in this mobile scenario despite complex underlying queries on large data sets.

6.2.3.4 Bringing In-Memory Technology to Desktop Applications

Since the introduction of reporting applications for personal computers, spreadsheet applications have been one of the most powerful tools for business applications. Business experts could now perform calculations interactively, which in the past had required repeated, manual computations or support from IT experts. It became possible to adjust the computations and see the results instantly. But there was one shortcoming: the input of data still had to be maintained by a dedicated input and data-cleansing center. Processing larger data sets required more capable computers managed by IT departments. In particular, the introduction of ERP, CRM, and Supplier Relationship Management (SRM) software systems running on centralized servers operated by IT personnel, created huge data volumes not available to business experts for analysis. To fulfill the analytic requirements, IT experts created reports running on the same servers. Later, BI systems were developed to allow analysis with increasing complexity. For further details, please refer to Sect. 7.1. While this IT-driven process ensured a high data quality, its disadvantages were inflexible and time-consuming operations to create or change these reports. Rather than providing

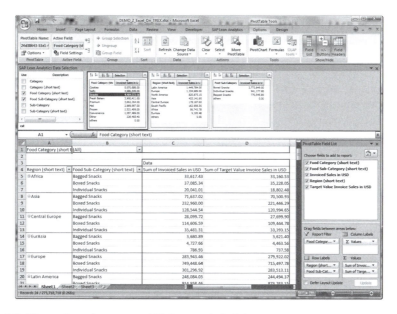

Fig. 6.6 Microsoft excel prototype: drill-down analytical data

predefined and inflexible user reports, from our point of view, an elegant way is to create software extensions for spreadsheet applications that access the database, ERP, or BI systems directly. As a result, the aforementioned gap can be narrowed, but a new problem became apparent: in spreadsheets, users expect results to be computed almost instantly, but transferring the data from the database to the spreadsheet is a slow, time-consuming process. Consequently, data are requested less frequently and the requested kind almost never changes.

From our point of view, a perfect combination requires three parts:

1. Easy-to-use spreadsheet software, which contains a rich functionality that supports the business user's creativity.
2. An extension of the spreadsheet software, which enables the access and analysis
   of large data volumes as easily as the spreadsheet itself. Ideally, the business
   user does not realize the difference between local and remote data storage.
3. A high-performance database, which offers response times comparable to
   spreadsheet calculations and a similar flexibility.

In summary, the goal is to combine the creativity of business users with the computational power of SanssouciDB.

The prototype shown in Fig. 6.6 shows Microsoft Excel as a frontend, exploiting the benefits of fast access to large volumes of data made possible by the use of SanssouciDB. The aim of the prototype is to show how to integrate high-speed in-memory data processing in office applications. The data is displayed and analyzed using the pivot table shown in the lower part of Fig. 6.6, which is expected by a business expert familiar with Microsoft Excel. The business expert can extend her/his analysis by including the data in formulas or add charts, as provided by Microsoft Excel. The data selection is done by an application add-in, which is visible in the upper part of Fig. 6.6, enclosed in a dark blue frame. Instead of a complex language or user interface, the user can interact with the application with the point-and-click paradigm and receives direct feedback. The business expert can search for measures and dimensions in the left panel. As soon as a dimension is selected, it is added to the form of a top k query. In this example, the first dimension was Food Category and the top ten food categories with the highest invoiced sales are shown. Each dimension is added in this way. At the same time, aggregated data is retrieved and added to the Excel Pivot table. Measures are added in the same way. In the given example, three dimensions Food Category, Region, and Food Subcategory and two measures Invoiced Sales and Target Value Invoice Sales were chosen. One or more values can be selected in the upper part of the prototype, for example, Food Category value Snacks, which filter selected data.

In this example, we were able to achieve data selection in sub-second response time even with data amounts of more than 275 million rows. This allows the business expert to model queries in real time with real data—a process that takes minutes or even longer in regular disk-based databases. Such scenarios are called agile modeling or self-service BI. The prototype shown was implemented without modifying Microsoft Excel. We used the Microsoft Visual Studio Tools for Office for accessing existing data interfaces. It was developed independently from any Microsoft product activities. This business example shows how a highly functional user interface, Microsoft Excel, is used to access data within the persistence layer. The orchestration layer is part of the Microsoft Excel add-in to enable the connection between data and user interface controls.

6.2.4 Combining Analytics and Textsearch

One of the major challenges for information workers today is to identify relevant information. The number of available sources of information is steadily increasing. For example, the Internet is a major source of digital information of all kinds. To access relevant information search engines such as Google, Bing, Yahoo, etc. are used. The challenge is to cope the variety of different document formats and to index the increasing amount of new and adapted information. Although internet search engines are known by all internet users, business data is far away from begin accessed in a comparable way.

In the following, we want to distinguish two categories of data that can contain information. Firstly, structured data which is any kind of data that is stored in a format, which is automatically processed by computers. Examples of structured data are ERP data stored in relational database tables, tree structures, arrays, etc. Secondly, unstructured data is the opposite of structured data, it cannot be processed automatically. Examples of unstructured data are all data that is available as raw documents, such as videos, photos etc. In addition, any kind of unformatted text, such as freely entered text in a text field, document, spreadsheet or database, are considered as unstructured data unless a data model for its interpretation is available, e.g. a possible semantic ontology.

In the following section, we share requirements for text search in enterprise applications with our focus on a medical application. We selected this example to highlight the fact that in-memory technology is not only useful in a special category of applications. In-memory technology can be a key-enabler applicable to any large data sets. Based on a cooperation with the hospital Charité—Universitätsmedizin Berlin, we investigated the following search capabilities.

6.2.5 Basic Types of Search

In the following, we distinguish the following basic search types.

- Exact search: Exact search is a crude tool from the early days of text search. The exact search performs a 1:1 matching of the search term and the identified text. This is applicable if the user knows that there is only a small set of selectable search terms, for example, in automatically generated documents. However, many documents that are generated by humans contain synonyms or spelling errors. As a result, the exact text search is only applicable for a small set of business data. For example, a medical doctor is looking for "carcinoma" and "adeno". Exact search will only find the exact match, but not "karzinom".
- Phrase search: In contrast to exact search, phrase search combines multiple words into a phrase and checks for its occurrence. As a result, phrase search involves the ordering of words within a phrase whereas exact search only focuses on the existence of all terms within a query. For example, if the medical doctor searches for "adeno carcinoma" the phrase search will only return results for the phrase, not for "adeno" or "carcinoma" or "carcinoma adeno".
- Linguistic search: Linguistic search is a powerful tool to specify search queries in natural language while a preprocessor derives relevant search statements. From the user's perspective, this is the easiest way to specify a search query. However, it requires a semantic analysis of the data to be searched as well as the query. This is a complex task that can be performed semi-automatically. For example the search for "Who is working at the Hasso Plattner Institute?" requires the identification of relevant terms in the query, such as "who", "working", and "Hasso Plattner Institute" as well as the mapping of these terms to corresponding data fields in the data, for example the database table `employees`.

6.2.6 Features for Enterprise Search

In the following section we focus on features that improve search results.

- Fuzzy search: The fuzzy search is an extension for any kind of the aforementioned search types. It handles a specified level of fuzziness in search queries automatically, e.g., typing errors. Fuzzy search blows up the pool of words that are searched for by inverting pairs of letters or scrambling them. With these methods additional words can be found that are stored in the wrong format in the data set to search in. This is very helpful if the data in the database was added by humans, because not only the people who type in the search term, also the people who entered the text you want to search on make mistakes. For example, if a medical doctor stores the results of a medical round he can choose between various descriptions for the same result, e.g., "carcinoma", "karzinom", "carzinoma", etc. Fuzzy search helps to identify them as relevant for the same search query.

- Synonym search: Natural language comes with the challenge that multiple words can have identical meanings. These synonyms can be used in various contexts, but the search query will contain only a single representation. For example, the medical abbreviation "ca." and "carcinoma" can be considered as synonyms. However, "ca." can also be the abbreviation for "circa". In other words, synonyms can have different meanings in different contexts. To keep track of them, abbreviations should be considered by their probability of occurring in the corresponding business context. To handle all these meanings you need an option to specify in which context a given meaning is valid.

- Ranking: A ranking of search results is needed to quantify the relevance of results for a certain query. In other words, ranking can be considered as a function $score(query, result) \rightarrow [0, 1]$ that returns a value in the interval $[0, 1]$. For example, in a tumor database the synonym "ca." for "carcinoma" gets a higher scoring than "circa" or "calcium".

- Did you mean? Another feature of a text search is providing the user a guess of what he could have meant, by providing a "Did you mean?" response. These guesses include the results of fuzzy or synonym search and replace the initial search query by a search query that results in more responses. For example, when a doctor wants to search through his diagnosis texts and he typed in "karcinom" it is valuable to see that the majority of search results contains "carcinoma" instead.

- Highlighting: After getting ranked search results, the context in which the search term occurs is important for the querying user. Therefore, a feature to highlight the individual relevance of a result set can be of use. Highlighting gives the querying user the option of showing the text he searched for in a specific format, e.g., underlined or bold. For example, a medical diagnosis may contain lots of text, if the doctor looks for the intensity of a radio therapy session a search for the unit "Gray" will be highlighted and the near context containing a short preceding and succeeding number of letters is returned. As a result, the intensity of the radio therapy session can be accessed in the search result without opening the diagnosis document.

- Automatic summary: The automatic summary of documents should give the user the option to decide individually whether the document is relevant or not. For example, the doctor does not need to open and read the complete diagnosis text, because all relevant information is extracted and summarized in the search result.

- Feature and entity extraction: A summarization is not the only information that can be extracted from a document. There is also the feature of entity extraction that extracts relevant keywords and names of entities from documents. This is comparable to tagging in online web blogs when certain entities are associated with a document. Especially when it is not clear what to search for, this automatically extracted information can help to add value the contents of the search result. For example, if a doctor is looking for tumor details the search results should contain only documents that are related to this topic even when the query does not contain the word "tumor". Entities in this context can be names of companies, individual persons or products; features can be medicine doses, number of patients, etc.

- Document clustering: In addition to document summaries, clusters of documents can provide a deeper insight. For example, when a diagnosis is related to the biopsy of the same tumor, these documents belong to the same patient and symptoms. As a result, this relationship is valuable for subsequent searches, e.g., for the tumor size or grade.
- Document classification: The category of a document is a piece of meta data that can be relevant for searches. For example, if the doctor is interested in drugs it is important where he found them, e.g., on prescriptions or chemotherapies. As a result, knowing the "type" of document helps to narrow the search result. In addition, it provides information about how to handle the search result.
- How in-memory-based search engines can support daily software engineering tasks: Another example of the combination of structural and textual data is source code querying. Software engineers are required to deal with a large amount of information about source code. Source code documents are of a dual nature: they are text containing information for developers and they have an explicit structure for compilers and other tools. Several representations for the structured information of source code exist: abstract syntax tree, call graph, data flow graph, and others. Although the questions developers ask about source code seem easy to formulate, the complex code structure often requires the writing of intricate queries. Developers use both lexical and structured information for queries.

An example of such a query is a search for all places where the variable "counter" is used in the condition of an IF statement. Another example is a search for compute statements that assign a value to a global variable. Those queries can be expressed as structural patterns of abstract syntax trees (ASTs). For example, the queries "find all function invocations which import a parameter named filename" or "find all reading accesses to a database table but000" are in essence AST patterns. If the developer does not know the exact identifier, additional semantic information should be added: a term dictionary, a translator between application-domain terms and implementation domain terms [132].

Our approach works as depicted in Fig. 6.7. Source code is parsed (1) and the resulting ASTs are stored into a central database (2). The database has a column-oriented layout and resides in main memory (3). Developers formulate their queries as sample code snippets (4), which are parsed into AST fragments and the corresponding XPath expressions describing those fragments are generated automatically (5). XPath queries are evaluated by a query engine and the result list contains links to source code documents as it would in any other search engine (6). The in-memory-based repository allows fast evaluation of structural queries and involves the semantic dimension of source code [131].

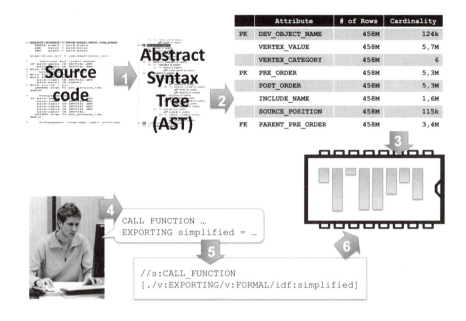

	Attribute	# of Rows	Cardinality
PK	DEV_OBJECT_NAME	458M	124k
	VERTEX_VALUE	458M	5,7M
	VERTEX_CATEGORY	458M	6
PK	PRE_ORDER	458M	5,3M
	POST_ORDER	458M	5,3M
	INCLUDE_NAME	458M	1,6M
	SOURCE_POSITION	458M	115k
FK	PARENT_PRE_ORDER	458M	3,4M

```
CALL FUNCTION …
EXPORTING simplified = …
```

```
//s:CALL_FUNCTION
[./v:EXPORTING/v:FORMAL/idf:simplified]
```

Fig. 6.7 Code analysis with in-memory source code repository

6.3 Conclusion

In this chapter, we described implementation techniques and general development guidelines that enable us to take full advantage of the performance improvements offered by in-memory technology. We showed how the speed-up gained by using an IMDB allows us to remove logical layers that were added between the DBMS and consuming applications to compensate for lack of DBMS performance. This in turn means we can do away with much of the costly transportation of data between the database layer and the application layer. The application stack itself becomes much leaner, both in terms of lines of code, and number of modules making maintenance much easier. We went on to describe how we re-implemented some of the enterprise applications introduced in Chap. 2 with an in-memory paradigm in mind and how completely new applications are enabled by this technology. Two major aspects characterize them: firstly, they involve the processing of large volumes of data, starting from millions up to billions of tuples. Secondly, their response time is expected to support the speed of thought concept mentioned in Chap. 1. In other words, these new kinds of applications are designed for interactive use rather than being executed as long-running background jobs. By providing the performance required to meet these criteria, SanssouciDB allows existing applications to be used in innovative and more productive ways. It also enables developers to create completely new applications that can help businesses run more effectively.

Chapter 7
Finally, A Real Business Intelligence System is at Hand

Abstract In this chapter, we offer our most important insights regarding operational and analytical systems from a business perspective. We describe how they can be unified to create a fast combined system. We also discuss how benchmarking can be changed to evaluate the unified system from both an operational and analytical processing perspective. As we saw earlier, it used to be possible to perform Business Intelligence (BI) tasks on a company's operational data. By the end of the 90s, however, this was no longer possible as data volumes had increased to such an extent that executing long-running analytical queries slowed systems down so much that they became unusable. BI solutions thus evolved over the years from the initial operational systems through to the current separation into operational and analytical domains. As we have seen, this separation causes a number of problems. Benchmarking systems have followed a similar trajectory, with current benchmarks measuring either operational or analytical performance, making it difficult to ascertain a systems true performance. With IMDB technology we now have the opportunity to once again reunite the operational and analytical domains. We can create BI solutions that analyze a company's up-to-the-minute data without the need to create expensive secondary analytical systems. We are also able to create unified benchmarks that give us a much more accurate view of the performance of the entire system. This chapter describes these two topics in detail. In Sect. 7.1, we cover the evolution of BI solutions from the initial operational systems through the separation into two domains, and then we give a recommendation regarding the unification of analytical and operational systems based on in-memory technology. In Sect. 7.2, we examine benchmarking across the operational and analytical domains.

7.1 Analytics on Operational Data

BI is a collection of applications and tools designed to collect and integrate company-wide data in one data store and to provide fast analytical, planning, forecasting and simulation capabilities. By way of introduction, a chronology of the eras and milestones in the evolution of BI systems is presented as historical background.

In this section we outline the stages of BI development, we discuss the technologies and concepts introduced to fulfill the requirements of business users, and we illustrate the drawbacks that arise from separating analytical from operational systems. We discuss technologies that enable changing BI and uniting operational and analytical systems and show what tomorrow's BI can look like.

7.1.1 Yesterday's Business Intelligence

At the end of the fifteenth century, Luca Pacioli, who is regarded as the Father of accounting, introduced the double-entry bookkeeping system with a debit and a credit side where each transaction was entered twice, once as a debit and once as a credit [130]. By using this method, merchants were able to analyze the financial status of their company.

Computer-based information systems that support decisions and related analysis tools were in use even before the emergence of relational database systems. The first prototypes of relational database systems were developed in the 1970s, and the first products were offered in the early 1980s. In Fig. 7.1, the development stages of analytical and operational systems are outlined along the timeline. The part below the time scale sketches the development of the operational systems. They have undergone continuous development, beginning in the 1960s and have become more sophisticated and heterogeneous over time. The different shapes of the three ERP systems shown illustrate the growing heterogeneity. Despite their heterogeneity, they interact with each other through standardized interfaces. The ever increasing amount of data from various and heterogeneous sources is visualized in Fig. 7.1 as oval star shapes of different sizes to emphasize their difference. Data from these sources has to be processed for daily operations. Factors that drove the development of operational systems were user interfaces and developments in storage technology. The introduction of monitors led to enterprise systems becoming accessible for more users and not only trained experts. The usage of disks instead of punch card systems with tape storage advanced the development of enterprise systems. Tapes still remain for backup and archiving purposes.

The upper part of Fig. 7.1 presents the analytical systems, which had to keep pace with the development of the operational systems, while having to cope with the changing analysis needs of users, for example, larger data sets being analyzed and more up-to-date data being requested. The current development whereby disks are being replaced by in-memory storage is an inflection point for operational and analytical systems of equal magnitude.

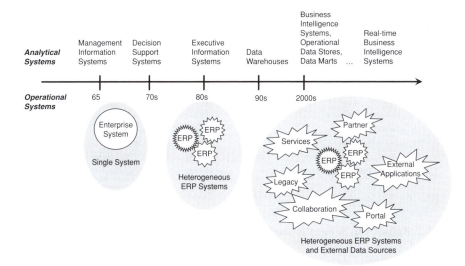

Fig. 7.1 Development stages of analytical and operational systems

7.1.1.1 Information Systems for Decision Making Before the 1990s

In 1965, the development of powerful mainframe systems, for example, the IBM System/360, enabled information retrieval from large data management systems [140]. The introduction of Management Information Systems (MISs) was intended to provide managers with structured reports to positively influence planning and control processes within companies. At that point, managers were not able to work directly with the data in the computer systems. Reports were usually created in periodic batch runs from which the manager had to filter relevant information after receiving a full report. Tools for structured problem solving and standardized methods, for example, drilling down, slicing and dicing, had not yet been introduced. MISs did not support the interactive analysis of data [22].

As a result, Decision Support Systems (DSSs) were developed in the 1970s to assist management interactively during the decision making process. This was made possible by introducing tools that helped explore information related to the problem domain. Spreadsheet applications are a typical example of DSSs that became widely used. At this time, spreadsheet applications were focused on single users exploring raw data to derive information. This was of limited use compared to a company-wide homogeneous data source. At the time, with spreadsheet applications, users were not yet able to collaborate on the same view of data.

Table 7.1 Characteristics of operational and analytical processing workloads

Operational Processing	Analytical Processing
Pre-determined queries	Ad-hoc queries
Simple queries	Complex queries
Highly selective query terms, small found sets	Low selectivity, large found sets
Inserts, updates and selects	Mainly selects
Real-time updates	Batch inserts and updates
Short transaction runtimes	Long transaction runtimes
Large number of concurrent users (1000+), large number of transactions	Small number of concurrent users (100+), few transactions

7.1.1.2 Reasons to Separate Analytical and Operational Systems

In the early 1990s, the separation of analytical processes from operational systems began, because analytical workloads had become more sophisticated and required additional computing resources. The users of analytical reporting, for example, required more data as a basis for reports to analyze entire business domains. Scanning large amounts of data for analytical purposes strained traditional operational systems at the costs of operational processing. The separation decreased load on the operational systems. As a result, analytical data was stored in a format that was prepared for analysis needs, and specialized optimizations of the data structures were introduced.

Operational and analytical processing exhibit different data access patterns that, along with database design, can be optimized. Table 7.1, based on [53], shows the characteristics of traditional operational and analytical applications as perceived in the 1990s. Operational applications typically execute the same predefined steps, which are parameterized according to current inputs. For example, a new order is completed for a customer listing products and their ordered quantities. Managers must be flexible regarding problem situations, and as a result their requirements for the underlying decision support system vary. Operational systems are often process-oriented, meaning that they are structured according to a company's processes and applications, such as order or invoice verification processing. Analytical systems are subject-oriented, clustered by analytical topics, such as sales analytics including customers, products, stores, and sales organizations. They cluster the main topics that managers need within a work context.

Compared to operational queries, analytical queries are relatively complex and may contain multiple grouping and sorting criteria as well as aggregations and joins. Operational systems are designed and optimized for single or a small number of updates, inserts, and modifications per transaction. They are designed for the mass throughput of many small transactions. In contrast, insertion of updated and new data into an analytical system is done via batch jobs that push in large amounts of data from operational systems.

Data is neither updated nor overwritten in an analytical system. This allows analysis tasks to include historical data as a basis for planning and simulation with the help of statistical methods. This multiplies the amount of data needed during analytical queries to a great extent. Consequently, we can deduce that there is a large difference between the execution time of analytical queries, which may take many hours, and operational ones, which typically complete within a few milliseconds.

7.1.2 Today's Business Intelligence

One of the reasons to setup and maintain BI systems in a separate environment is the need to integrate data from many different sources. For any given company, this data can be in the form of stand-alone spreadsheets, other documents and external information from the Internet, in addition to their ERP, CRM, SRM, or other operational back-end systems (Fig. 7.2).

A company's analytical data may be stored by software systems from different vendors or in different software versions and may have no common structure or semantics. The latter can be derived from metadata. For example, the domain type metadata of a data element that stores numbers can define, whether it is weight or volume information. Using data from different sources requires a consolidation of metadata and data. This consolidation is done during the ETL process.

7.1.2.1 Data Mart, Data Warehouse and Operational Data Store

The typical components in the BI landscape are data marts, data warehouses, and operational data stores. Data marts contain only data from a certain business area like sales or production. They structure data into clusters that provide the most relevant information for specific user groups. Analyzing cross-domain or company- wide information requires the use of a data warehouse. If data marts are mistakenly used for cross-domain analytics, more data marts will have to be introduced to satisfy new analysis requirements. The introduction of more data marts can rapidly lead to a large collection of data marts without common semantics, particularly if the design is not driven by a top-down analysis of the business. Instead of that collection of data marts, a data warehouse should provide company-wide and cross- domain analytics. Data warehouses contain summary data prepared for analyses in multiple dimensions. A centrally coordinated organizational unit of a company should own the data warehouse [83]. The goal of this unit is to achieve a balanced representation of all domain subjects of that company. The data warehouse should be constructed iteratively to avoid many years of building it before it can be used. The idea behind running a data warehouse is that it provides the ability to analyze all the data of a company in a single system. The reality, however, is different: large companies often have multiple data warehouses. The largest data warehouses triple in size approximately every two years [186]. Reasons for this increase are as follows:

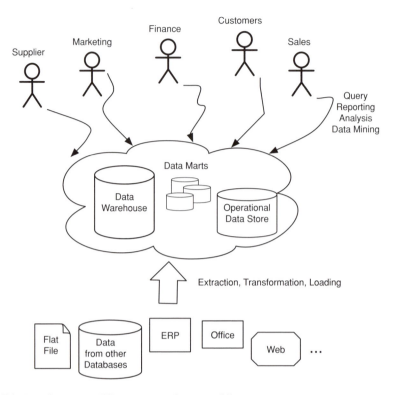

Fig. 7.2   Data from many different sources for many different consumers

- Redundancy in data storage,
- Storage of data at an increasingly detailed level,
- Inclusion of data concerning more and more subjects and domains,
- Storage of new types of information, for example, unstructured textural data, audio, video, and sensor data,
- Increased data retention period.

If the size of the database becomes greater than the amount of main memory available then data partitioning and data aging (Sect. 5.2) become important. To summarize briefly, there is a need for data warehouse systems to consolidate heterogeneous data and to make this data available as high-quality corporate information. The rapidly increasing data volume requires dedicated high-performance systems. Operational Data Stores (ODSs) represent the information of one source system on the same level of detail and they provide different levels of data latency from almost instantaneous to daily updates that are pulled from the source system [84]. Data from ODSs can be migrated and transformed into data marts and the data warehouse to avoid pulling the same data from operational systems several times. ODSs contain only a small amount of data for a limited time frame to allow short- term analysis and support so-called operational reporting; that is, detailed reporting with up-to-the second accuracy [85].

7.1.3 Drawbacks of Separating Analytics from Daily Operations

The strict differentiation between operational and analytical applications is a simplified view of reality. In each domain, many applications exist that show characteristics of the other domain (Chap. 2). As a result, a black-and-white picture is often not correct. For example, data stored in a data warehouse is not always summary data. Line items may also be stored in the data warehouse to allow analyses with a high granularity, that is, the same granularity as is encountered in the operational systems.

The data warehouse is not just the end of the data processing chain. The integration of data warehouses with, for example, ERP, CRM, or SCM applications to access analytical data is beneficial, and data in the data warehouse may also serve as a source of information for operational systems. Data in the analytical system landscape is not exclusively historical data. Up-to-date data for near real- time analysis often has to be provided and ODSs with different data latency types have been introduced to fit this demand. A consequence of separating analytical from operational systems is the need to constantly replicate data. The ETL process covers this replication:

- *Extraction:* Process step that periodically pulls data from various data sources and copies it into a staging area in the analytical system.
- *Transformation:* Process step that includes reformatting, correcting data to remove errors and inconsistencies, and enriching data with external data. Computations such as duplicate removal and cleansing can be executed while copying data into the final analytical structures that are prepared for query consumption. Examples of transformation operations carried out in the data warehouse environment are shown in Table 7.2.
- *Loading:* Once transformation is done, the high quality data is loaded into the destination data center and made available for business intelligence operations. The common data model ensures that BI users access the same data definitions when they perform reporting and analysis functions.

Companies often regard ETL as the most difficult part of BI. The reason for these difficulties is the heterogeneous nature of data that comes from diverse data sources, including RDBMSs, flat files, Microsoft Excel, packaged applications and Extensible Markup Language (XML) documents. Since decisions may need to be made based on near-real-time data, data has to be loaded more rapidly and more frequently into the storage used by the analytical applications in order to have up-to-date information. 24×7 availability and shrinking batch windows for data warehouses that span regions and continents leave no dedicated downtime for the system to execute extracts and loads. The data volumes are very large and data warehouses with terabytes of data are common. Another issue is that transformations, for example, cleansing, can be very complex and time consuming.

Table 7.2 Transformation steps during ETL

Transformation	Description/Example
Missing values	Not all source systems are able to provide data for all fields. In this case, default values or values retrieved from look-up tables are used to fill in any missing field values.
Cleansing, duplicate removal, harmonization	This operation makes sure that data stored in the data warehouse is correct and complete. As an example, gender should always be M or F. If source systems use differing representations for gender, this data has to be harmonized.
Referential integrity	Transformations ensure that referenced data actually exists, for example, the master data tuple for a product referenced in a sales order.
Reformatting	Establish a consistent representation for all data items of the same domain type; for example, reformatting the date value of different source systems: change 01/31/2010 to 2010-01-31.
Type conversions	Change data types of values from different source systems according to the most useful type for reporting, for example, character to integer, string to packed decimal.
Character set conversions	The conversion of ASCII to UTF8 is an example.

Overall, ETL increases the TCO. Many BI projects are not realized because the TCO of the solution would be higher than the expected benefit. See Sect. 1.3 for more information regarding TCO.

7.1.4 Dedicated Database Designs for Analytical Systems

Some of the first dedicated analytical systems were called Multidimensional OLAP (MOLAP) processors because they used optimized multidimensional array storage. Unfortunately, their practical usage was limited to relatively few dimensions while many business scenarios work on dozens of measures and up to 100 dimensions.

Relational OLAP (ROLAP) designs were created that used existing relational databases as storage. Their design allowed the system to handle large numbers of measures and dimensions and large data volumes. Database schemas in an operational environment are highly normalized to increase write performance and avoid update anomalies. ROLAP systems often use denormalized schemas to avoid join operations. Based on relational database storage the star schema design has become established as the standard representation of data for analytical purposes. The key elements within the star schema are measures, dimensions and facts.

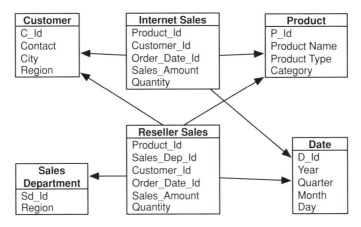

Fig. 7.3 Star schema example for internet and reseller sales

- Measures are all attributes that can be aggregated. The definition of a measure includes the function for aggregation: for example, sum or average.
- Dimensions hold all other attributes that cannot be aggregated, but can be used as filters describing the business event: for example, year, country, product name, or department.
- Facts are tuples of measures and references to dimensions. A fact represents a concrete business event: for example, sales of a product. Each tuple represents a specific relation of measure and dimension values.

A typical example for a star schema for Internet and reseller sales is shown in Fig. 7.3. It includes two fact tables `Internet Sales` and `Reseller Sales`, which hold the Sales Amount and Quantity measures that can be aggregated along data in the four dimensions `Customer`, `Product`, `Date`, and `Sales Department`. Since internet sales are direct sales, no sales department is involved here.

The fact table is by far the largest table in the star schema. It is updated frequently with tuples created for each business event. According to the level of granularity, which is defined by the maximum resolution in the dimensions, the data of each business transaction finds its representation in the fact table. In our example, the finest level of granularity is orders per customer, product and region per day. Assume a customer has two Internet orders as follows:

```
Order 1024, Customer 54, 2010-09-26, Sales key: 4
    200 cases Clear Mountain Water
    500 boxes Spicy Bacon Chips
Order 2397, Customer 54, 2010-09-26, Sales key: 4
    750 boxes Spicy Bacon Chips
```

The following lines are added to the Internet sales fact table in addition to the appropriate identifiers for the dimensions per line:

```
200 cases Clear Mountain Water
```

```
1250 boxes Spicy Bacon Chips
```

The information that the initial order process was fragmented into several orders on the same day would be lost in this case, as the highest temporal granularity is set to day-level.

Details within dimensions represent master data, which is updated less frequently. For example, the details of products in the product dimension are rarely updated compared to the sales of this product. The size of these tables is much smaller than that of the fact table, as products are usually sold more than once.

Measures have to be part of the fact table so that aggregations only have to be performed on the central fact table. Before aggregating, dimensions are filtered according to conditions given in the query. Almost all queries contain restrictions; for example, sales of a certain product during this year. These restrictions are mostly applied to dimension tables and can be propagated from the outside of the star schema to the central fact table. In addition, one-to-many relations greatly simplify the join complexity and are a key to the superior read performance of this schema design.

Building a star schema to satisfy just one report is not feasible since many reports overlap with regard to their requirements concerning measures and dimension attributes. As a consequence a star, also called a cube, covers several reports. The cube introduces redundancy that grows exponentially with the number of dimensions. All dimension combinations that contain no valid aggregate data in the fact table are filled with null values that have to be stored. Our experience shows that even if the cube is compressed, it may need more storage space than the operational data it is based on. Building up a new cube from scratch is a lot of work, as data has to be extracted from the source systems and transformed into the final cube structure. It can be several weeks before the data is available for reporting, which is one reason why cubes usually contain more dimensions and measures than are initially needed so they can cater for possible future reporting needs. Updating an existing cube to provide additional data for a new report is also work intensive. As a consequence cubes are often much larger than necessary, placing overhead on the reports using them.

Many star schema implementations use shorter Surrogate Identifiers (SIDs) [92] instead of referencing the primary keys of the actual data set to create a fact-dimension relationship. They are integers and their usage results in a reduction of the required storage space for the fact table, while incurring a comparatively small overhead in the dimension tables to store the additional SID. SIDs simplify the joins of the fact table to the dimension tables and thereby improve join performance. An even more advanced realization of this idea is the usage of column-oriented storage and compression as provided in, for example, SAP Business Warehouse Accelerator [153] and ParAccel [133]. SAP BusinessObjects Explorer [152] on top of in-memory technology is an example of a product that provides an easy-to-use interface to interactively browse and analyze large amounts of data in an intuitive way using point and click interactions. Through simple user interactions it allows drilling down, rolling up, as well as slicing and dicing (aggregating along multiple dimensions) according to the user's access rights.

The snowflake schema is a variation of the star schema and is used to reduce the redundancy in dimensions, introduced, for example, through hierarchies that are encoded within one dimension. In the example shown in Fig. 7.3, the Category of the Product dimension is a candidate to be extracted into its own table. The snowflake optimization introduces further joins, which hampers query performance. The use of the snowflake schema is not generally recommended in a data warehouse, where the main reason for its existence is maximum query performance [139].

7.1.5 Analytics and Query Languages

Almost any business query must process hierarchical data. Frequently, the query contains a time frame using the time dimension, which is composed of the well-known hierarchy of, for example, years, quarters, months, and days. Other examples are regions or countries, organizational units, profit centers, and product groups. Not all of them can be transformed into uniform layers of dimensions, but may be irregular, for example, organizational hierarchies that contain different roles on the same layer. The hierarchical structure may be time dependent. Considering an organizational hierarchy example, the monthly sales report of a sales unit should reflect the organizational changes in the unit, for example, employees switching positions. Hierarchies, such as organizational hierarchies, can change frequently, and in many cases a record of these changes is needed for historical snapshot reporting. For this reason hierarchical data should be stored in its own structure and not in the fact or dimension table.

Business queries quite often require hierarchical set operations and features not available in relational database languages like standard SQL, or they require complex statements that interweave multiple dimensions. Another aspect of business queries is the way they interpret the data. Often, a multidimensional view instead of a relational one is required. An example query is showing the combined internet and reseller sales amounts for calendar year 2010 for the top ten products of calendar year 2009 based on combined sales during that year. Transforming this query into SQL requires several distinct grouping sets, making the query very complicated.

Commonly performed computations for analytics like cumulative aggregates, or determining a ranking of the top ten selling products, are very difficult using SQL [134]. To cope with this query complexity, a multi-dimensional data structure based on the star schema was developed. And particularly for multidimensional analytics, the query language MDX inspired by SQL was developed and became the de facto standard. This query language contains special functions for analytics:

- A notion of time semantics and computation, for example, monthly sales and three-month rolling average.

Table 7.3 Result table of complex computations of sales amount (in $) on multi-dimensional data

Product category	Reseller	Internet	Combined
All Products	80,450,596.98	29,358,677.22	109,809,274.20
Accessories	571,297.93	700,759.96	1,272,057.89
Bikes	66,302,381.56	28,318,144.65	94,620,526.21
Clothing	1,777,840.84	339,772.61	2,117,613.45
Components	11,799,076.66	–	11,799,076.66
Percent Bikes	82.41 %	96.46 %	86.17 %

- Computation of measures related to dimensional hierarchies as in sales amount of each product and percent of parent product level or category.
- Complex computations on multidimensional data and which require a specific order of computations to be correct. Our example here is percent sales of bikes for reseller, Internet and combined sales. Note, the result shown in Table 7.3, where the required computation for combined sales is not in the displayed result set but has to be done during aggregation.

These were just some examples of business queries run by commercial analytical systems. More complex queries are often necessary. Some of them can be transformed into large, complex SQL statements. Many operations, however, require reading the data from the relational store and processing it in an OLAP engine outside the database. In the case of separated database and OLAP processor systems, this causes large amounts of data to be transferred and slows down the response times substantially. It is important to perform such computations within or close to the database to ensure optimal performance.

In SQL, all joins to create the traversal through hierarchies, which means selecting data from different tables, have to be modeled manually. This is a fundamental difference to MDX statements, where metadata concerning the join conditions of tables within hierarchies is explicitly stored in the database and does not need to be modeled redundantly in each statement. Thus, MDX statements can be significantly shorter and easier to comprehend for complex analytical queries.

7.1.6 Enablers for Changing Business Intelligence

As data volumes grew, the limitations of MOLAP and ROLAP technologies became more apparent. Building and maintaining data warehouses in a separate BI landscape to compensate for technical deficiencies of operational systems is no longer necessary because in-memory databases will empower operational systems to provide complex analytical functionality. Performance and scalability issues can be addressed by a complete redesign exploiting in-memory technology, column-oriented storage, compression, shared-nothing, and multi-core hardware architectures. We further discuss some of the concepts and technologies that are changing the BI domain:

- The fast access times that in-memory data storage enables, especially in combination with lightweight compression, become interesting if all company data can be held in main memory. In analytical scenarios on a disk-based system, where multiple users concurrently analyze massive amounts of data within an overlapping data set to create their reports, the number of disk seeks increases drastically and introduces a large penalty on the query response times. In an in-memory system the number of random accesses grows, but the penalty is much smaller due to the faster random access of main memory storage. Sequential access is even faster, but it remains to be proven if a sequential execution of multiple parallel requests is superior in such a scenario.

- Column-oriented data storage not only facilitates compression but also is optimized for OLAP access patterns. OLAP actions typically touch only a small set of columns, for example, a filter and an aggregation. Not all fields, for example, of a large fact table, are needed. In column-oriented storage, the columns that are not required in a query are simply not read, as opposed to row-oriented storage where unused columns may clutter the cache lines if they are next to the values of a used column (Sect. 4.4).

- During Extract, Load, and Transform (ELT), compared to ETL, data is extracted and loaded first into the data warehouse in a highly parallel way without any (complex) transformations. In a subsequent step, data may be transformed if necessary. An offline copy of the entire data set is created and the extraction and loading step can be decoupled from transformation.

- Stream databases are designed in a way to avoid unnecessary storage operations. Using in-memory database technology for this purpose seems to be quite promising, especially when additional historical analyses are required. For example, algorithmic trading in financial services, fraud protection, and RFID event processing require processing a large data volume in a very short time (approaching milliseconds in algorithmic trading).

- Empower business professionals to do their analysis without extensive IT support. While IT support staff, data warehouses, and well-defined processes are still required to ensure a sufficient quality of data, business professionals complain about waiting periods until change requests are implemented in their BI tools. Some companies have developed local, in-memory based tools, such as QlikView [141]. Others integrate an in-memory database in the local spreadsheet application and relate it to the data warehouse run by IT, for example, Microsoft PowerPivot [114].

While business intelligence has become a must have for companies and is perceived as an established technology, we observe that the focus of business users has shifted to real-time analytics. Companies are attempting to overcome the delay caused by ETL and implement systems allowing near real-time access to more of their existing or new business data. This necessitates a shift from analyzing historical data to real- time analysis of current data for business decisions. One example is the increasing number of so-called embedded analytics applications, where operational style data is combined with analytical views of historical and current data to support decisions. Integrating the results of analytical reports with operational data and providing that integrated data set for further analyses is the goal of a special type of ODS [84]. Calculations that include historical and current data in combination require even more computational resources than real-time or historical analysis and further strengthen the importance of the in-memory database paradigm for analytics.

7.1.7 Tomorrow's Business Intelligence

In the early days of BI, running queries was only possible for IT experts. The tremendous increase in available computational power and main memory has allowed us to think about a totally different approach: the design of systems that empower business users to define and run queries on their own. This is sometimes called self-service BI.

An important behavioral aspect is the shift from push to pull: people should get information whenever they want and on any device [74]. For example, a sales person can retrieve real-time data about a customer, including BI data, instantly on his smart phone. Another example could be the access of dunning functionality from a mobile device. This enables a salesperson to run a dunning report while on the road and to visit a customer with outstanding payments if she or he is in the area. These examples emphasize the importance of sub-second response time applications driven by in-memory database technology. The amount of data transferred to mobile devices and the computational requirements of the applications for the mobile devices have to be optimized, given limited processing power and connection bandwidths.

As explained previously, the principal reasons to separate analytical from operational systems were technical. An exception was the need to consolidate complex, heterogeneous system landscapes. As a result of the technological developments in recent years, many technical problems have been solved. We propose that BI using operational data could be once again performed on the operational system.

In-memory databases using column-oriented and row-oriented storage, allow both operational and analytical workloads to be processed at the same time in the same system. The need for redundant data structures, like the cubes of the star schema, does not exist anymore and views or virtual cubes can replace them enabling analytical applications on the presentation layer to access a virtually unified data store. On the data management layer, the data should be stored with as little redundancy as possible. The most prominent approach is to store data in normalized fashion, e.g. third normal form, as has long been proposed to avoid update, insert, or delete anomalies. However, this has so far been traded against speeding up operations, e.g., by adding master data to the transactional tables and thus reducing the number of joins. Yet, simplification of applications through using a normalized data set and not having to deal with manually handling the anomalies in each application is far more desirable than reducing the number of joins, especially in the case of in-memory databases with optimized join algorithms. Calculations defined in the transformation part of the former ETL process are moved to query execution time.

Many advantages of this architecture have been mentioned in previous sections. In summary, we can say that it enables the instant availability of the entire data set for flexible reporting and reduces the TCO and complexity of the solution. Moreover, this approach fulfills many of the ideas mentioned in the context of BI 2.0 discussions, where BI is "becoming proactive, real-time, operational, integrated with business processes, and extends beyond the boundaries of the organization" [143].

Planning applications show characteristics of operational as well as analytical workloads. A large and diverse family of business planning processes and methods exists—for example, demand planning or distribution planning. They all share at least one common requirement: the planning process must be finished before its results become outdated. This process includes many alternating steps of plan changes and aggregation. For example, the change of the planned overall sales volume has to be distributed to the lowest layer containing the detailed data. The updated detail data is aggregated in various ways in subsequent steps to analyze and adjust the new planned data, for example, for each product, and region. Large numbers of tuples are updated and aggregated in many steps, which in the past took too long to do in an online process. In-memory solutions provide planning capabilities within seconds. These kinds of applications with immediate response times not only improve the user experience, but they have the potential to enable completely new business applications, which have been impossible so far.

7.2 How to Evaluate Databases After the Game has Changed

Benchmarking is the systematic comparison among one or more systems based on well-defined test settings. In the context of this book, we will take a particular look at benchmarking database systems used for both operational and analytical data processing within business environments. Recent developments in the analytical domain show that databases that primarily store data in main memory are gaining ground. Column-oriented data storage is becoming the standard for analytical systems; see, for example, SAP Business Warehouse Accelerator [153], ParAccel Analytic Database [133], or Exasol EXASolution 2.0 [49]. The question of whether these databases, which have been developed for the analytical domain, are also able to handle operational or mixed workloads, and could thereby reverse the separation of analytical and operational systems, is still in dispute.

In order to allow us to answer a question like the one above, that is, assessing the capabilities of databases within a specific domain, benchmarks represent a widely accepted method for comparison and evaluation. Benchmarks represent the consensus of what to measure and how to measure [62]. As a result, benchmark results are comparable. With regard to the above questions and considering the specific characteristics of enterprise applications of both domains (Chap. 2), existing benchmarks

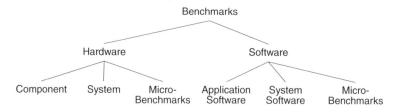

Fig. 7.4 Benchmarking categories

have to be adapted or new benchmarks have to be introduced to evaluate database systems for a mixed operational and analytical processing environment. We will outline additional parameters that need to be considered, especially when benchmarking across the two domains and give an overview for a benchmark scenario that is meaningful within a business context and that provides a basis to assess characteristic actions of both domains.

7.2.1 Benchmarks in Enterprise Computing

Benchmarking computer systems can be separated into two main categories: hardware benchmarking and software benchmarking, see Fig. 7.4 for an overview of benchmarking categories. Hardware benchmarks are categorized as component and system benchmarks. Component benchmarks assess the performance of specific system components. The Standard Performance Evaluation Corporation (SPEC) provides widely used benchmarks for this purpose. These include CPU2006 [29] for CPUs and SPECviewperf [30] for graphics. In contrast, system benchmarks assess the performance of an entire hardware system.

Software benchmarks can be categorized as application software benchmarks, system software benchmarks. An example of application software benchmarking is assessing the performance and functional completeness of different word-processing applications. System software benchmarks target all software that supports the running of user software, for example, operating systems, virtualization software (see SPECvirt [34]), or web and mail servers (see SPECweb [32] and SPECmail [31]). For enterprise computing the most relevant areas are application and system software benchmarks. Micro-benchmarks exist for hardware as well as software. They measure the performance of small and specific parts of larger systems and are composed individually.

Application software benchmarks for business applications have been developed, for example, by SAP [154] and hardware partners, Oracle [127], or the SPEC, for example, SPECjEnterprise2010 [33]. Such benchmarks help customers to find the appropriate hardware configuration for their specific IT solutions and they help software and hardware vendors to improve their products according to enterprise needs. They measure the entire application stack of a given configuration and the results are insights into the end-to-end performance of the system, including the database, as if it was a real customer system.

Database system benchmarks model the behavior of a set of applications, which belongs to a certain business domain, and they emulate the workload behavior typical for that domain. They directly evaluate the performance of a given database system. For the two domains, operational and analytical processing, the database benchmarks of the TPC have become the standard database benchmarks and are widely used. TPC's currently active benchmarks for operational processing are TPC-C [175] and its successor TPC-E [176]. For analytical processing TPC-H [177] is the current standard and TPC-DS [138], a new benchmark for decision support systems, is now available. The star schema benchmark [125] is a variant of TPC-H. It derives a pure star schema from the schema layout of TPC-H to evaluate the performance of data management systems built for pure data warehouse environments.

Systems can be compared to each other with respect to different points of view: performance, functionality, TCO (Sect. 1.3), and energy consumption. In database system benchmarking the basic measuring unit to compare different hardware and software configurations is transactions per time unit. Depending on the benchmark, a transaction in this context can be a sequence of several actions: a simple example for a sales order transaction shows that ordering a product may consist of retrieving the details for the chosen product, checking its availability, and storing the order data if the customer chooses to buy the product. To consider the monetary aspect, TPC takes into account the price of the hardware of the evaluated system, including three years of maintenance support, for example, online storage for the database and necessary system software as well as additional products to create, operate, administer, and maintain the future applications. The energy specification [178] of the TPC offers rules and a methodology to report an energy metric in the TPC benchmarks. The energy consumption of typical components within the business information technology environment, for example, servers and disk systems, is measured.

7.2.2 Changed Benchmark Requirements for a Mixed Workload

In the following section, we focus on database system benchmarking and especially on pointing out the changed requirements for database benchmarks with mixed operational and analytical workloads. Such database systems dissolve the currently existing boundary between the operational and analytical domains.

Since operational and analytical benchmarking have been treated as separate domains so far, only limited statements can be made concerning the ability of database systems to handle a mixed workload using the existing benchmarks. Simply running an operational processing and analytical processing benchmark in parallel can lead to an incomplete picture of the actual performance of a system that handles a mixed workload. In this case, only the effects of hardware resource contention, that is, the interference of multiple processes running concurrently, are measured. The problem is that each benchmark still runs on its own dedicated data set. Conflicts arising from data access to the same data set are of particular interest when analyzing and optimizing systems for a mixed workload. New means are required to evaluate and compare new database systems integrating the mixed workload, while covering desired business scenarios.

Variations within the data set and workload are the key factors that influence the performance of database systems. Appropriate parameters are modeled in existing benchmarks to quantify the impact of the variation of the data set and the workload on the database system under test that can then be adapted and optimized accordingly. These parameters are load and data set size. Load is based on the number of concurrent clients and requests. Both parameters generally appear in existing benchmarks.

For operational and analytical processing benchmarks, standard database designs are used as the basis to model the data. The underlying database design of operational and analytical applications is very different. Therefore, another parameter needs to be taken into account: the database design itself (logical and physical). Operational data schemas are optimized mainly for the efficient recording of business transactions and single tuple access. As a consequence, high levels of redundancy are avoided preventing data inconsistencies and update dependencies [147]. Additional structures for optimized read performance, for example, indices, materialized views, or precomputed aggregates, introduce redundancy and add overhead to the insertion of new data. Analytical data schemas are query-centric and optimized for the aggregation of mass data [12]. In a database system optimized for a mixed workload, these diverging tendencies have to be harmonized to produce a single source of truth. Database design as a benchmark parameter finds no representation in existing database benchmarking efforts, but is vitally important in a combined operational and analytical scenario.

The database design has to be tuned for each specific database implementation to achieve the best results. It depends on the actual mix of the different workloads, meaning the ratio between operational and analytical style queries. No general database design exists that is optimal for all mixes of workloads. If, for example, the workload is mainly composed of analytical style queries, the database design may result in a more analytical optimized version, trading decreased join complexity for increased redundancy of data.

Workload mix does not exist as a parameter in current TPC benchmarks. They provide one standard mix of the given benchmark transactions to which the actual benchmark run adheres. TPC's operational web e-Commerce benchmark TPC-W [174], which has been marked obsolete since 2005, explicitly models various mixes of the benchmark transactions (browsing, shopping and ordering) in order to reproduce diverse user behavior. Workload mix, however, clearly becomes important when the operational processing and analytical processing domains are united.

In summary, load and data set size are widely accepted parameters that remain relevant for a benchmark that unites the two domains. Database design and workload mix are two new parameters that have to be introduced to such a benchmark to cover the diversity that exists b etween these domains.

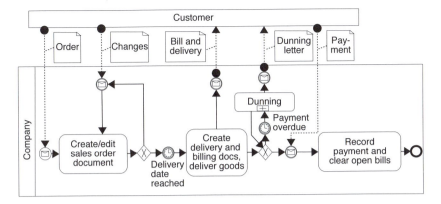

Fig. 7.5 Order-to-cash process

7.2.3 A New Benchmark for Daily Operations and Analytics

This section provides an overview of a benchmark that models a mixed workload and incorporates the characteristics mentioned above: the Combined Benchmark for Transactions and Reporting (CBTR) [15]. CBTR contains the previously mentioned parameters and is based on real business data. CBTR follows the order- to-cash scenario (see below) and it focuses on the management and processing of sales orders and their financial follow-up tasks, such as receipt of incoming payments, from the operational perspective as well as corresponding analytical functions, for example, organization effectiveness, shipping performance, and accounts receivable cash flow.

7.2.3.1 Order-to-Cash Data Entities

In detail, the process underlying the order-to-cash scenario consists of several steps: The creation of a sales order, the delivery of ordered items, billing delivered items and recording corresponding incoming payments. Figure 7.5, based on [16], provides an overview of these steps and shows the interaction of a co mpany with a customer.

Naturally, the database requires the appropriate data to support the process steps we described above. For the reader, to get a sense of the data involved in the process, we give a simplified overview of these entities involved and their relationships, including an excerpt of the most important attributes, see Fig. 7.6 based on [15].

The master data needed in the order-to-cash scenario is shown in gray shading. Master data does not change frequently during business transactions. In our scenario, master data includes customer contact information, customer contract data, product data (for example, description, weight and volume information), and sales organizational information (for example, products for sale and responsibility for certain customers). All other entities contain the operational data of the order-to-cash

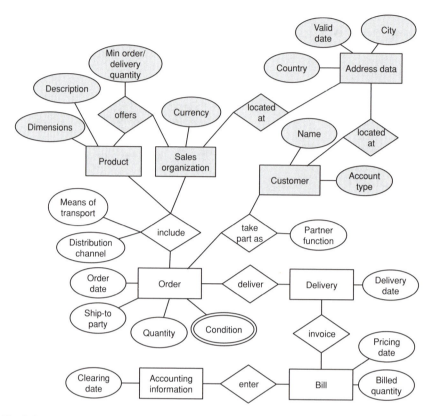

Fig. 7.6 Conceptual database schema

scenario. They include customer orders for specific products, deliveries of ordered products to customers, customer billing and records of the accounting information when bills are paid. The entities containing operational data change with each business transaction. They are connected to each other through references that are created during inserts and updates.

7.2.3.2 The Transactions and Analytical Queries

The operational side of CBTR emulates the most frequently used business functionality within the order-to-cash scenario. According to their data access type, the transactions can be distinguished into read-only access and mixed access. Mixed access refers to transactions that are composed of several actions and do not solely write to the database, but also contain read access parts. Transactions with mixed access in CBTR include the following:

- *Creation of new sales orders:* This transaction, which includes adding line items, customer information, desired products, quantities and the negotiated delivery date, and forms the foundation for later actions. Customer information is selected from existing master data in the database. The same process is applied to product details that are needed within the sales order.
- *Creation of deliveries based on sales orders:* This transaction checks if all items, which are going to be delivered, have been ordered, and are shipped to the same receiving party. Deliveries may cover several orders by one customer to the same destination or parts of existing orders.
- *Shipping:* After a certain amount of time, and according to customer contracts, sales order items are shipped. This has to be recorded in the database with the shipping date, the shipped items and a reference to the sales order in order to maintain a link throughout the process.
- *Creation of billing documents:* This action checks if all referenced order line items have been shipped before issuing the bill. Creating billing documents triggers the creation of associated documents in the accounting system, where the items are marked as open. Open line items are those items that have been delivered, but not yet paid.
- *Recording of incoming payments and clearing open line items:* This transaction pulls all open line items from the database and updates those that are referenced in an incoming payment document to clear them.

The read-only part of the operational workload can be categorized into detail lookups and range lookups. Detail lookup actions include displaying sales order details, customer details, and product details. The access path consists of the selection by key. Brief range lookup actions include the following actions with reference to a certain time frame: Showing all sales orders and their items and displaying unpaid orders.

Only a brief period of time is of interest for most operations in transaction processing, for example, orders or due invoices within the last month. This lookup is a typical operational reporting query. Operational reporting queries are analytical style queries for decisions based on recent data. Only a short time frame, for example, one month, is considered and, as a result, a relatively small set of data is analyzed compared to reporting queries. Expansion of the time frame is one dimension that leads to a larger set of data that is evaluated for the result, thereby increasingly exposing analytical processing characteristics.

The analytical queries reflect typical query types encountered in a real enterprise environment. They include the following, listed in the order of increased join and computational complexity:

- The daily flash query identifies today's sales revenues, grouped, for example, by customer, product, or region.
- The customer payment behavior query analyzes the time from the billing of orders and the receipt of incoming payments, which is an important factor for the liquidity of a company. It gives an overview of the general customer payment behavior. It is also known as days sales outstanding or days receivables.

- The average order processing time query that determines the entire order processing time starting from an incoming order until the last item has been delivered to the customer within a given time frame. It helps to monitor order processing and ensure customer satisfaction.
- The sales order completion query is included because the percentage of sales orders that have been shipped completely and on time within a given time frame is of interest to ensure service levels.

To achieve a realistic workload, the detailed mixed access operational queries of a real enterprise system are incorporated into CBTR. With regard to read-only operational and analytical processing, the workload mix is variable, as mentioned above, to measure its impact on the operational business transactions.

To summarize, the latest advances in hardware and in-memory database management have led to the conclusion that upcoming systems will be able to process mixed workloads. With the introduction of new systems, benchmarking has to evolve. Compared to dedicated operational and analytical processing benchmarks, a benchmark for mixed operational and analytical processing workloads needs to take into account the differences that result from the optimization for the sole workloads. These differences are mirrored in variations of the database design to a large extent. Mixing the workloads on the one hand and varying the database design on the other hand are all additional parameters that have to be observed. A benchmark for mixed workloads needs to provide a scenario that includes a relevant amount of data to allow the emulation of a realistic workload for operational as well as analytical processing, by recreating characteristic actions.

As with existing benchmarks, a new benchmark for combined operational and analytical systems can help with the execution of two important tasks: creation of systems and comparison of systems. Creation of systems comprises of finding an optimal design and configuration for a data management layer given the specific needs of a company. Here all parameters can be tuned to a company's needs. For comparison of systems, as in existing benchmarks, discrete points on all parameter scales have to be defined to keep the benchmark results comparable across different products.

7.3 Conclusion

In this chapter, we revisited the separation of operational and analytical processing, taking a more business-oriented view on the problem. Our detailed look included the evolution of BI systems from their earliest incarnations, through the splitting of operational processing and analytical systems due to rising data volumes, right up to the separate operational and analytical systems that exist today. We also reiterated our view that the performance advantages offered by in-memory technology will allow us to process and analyze the large data volumes present in current enterprise applications in a single system.

As noted previously, the reasons for splitting enterprise applications into operational and analytical systems were performance and the ability to integrate data from different sources in a single data store for analytics. It is vital that we find a means to assess the reunited system to ensure that it shows superior performance vis-à-vis the separate systems as well as enabling the integration of data from several systems for analytics, for example, on the fly, during query processing.

Traditional ways of assessing the performance of enterprise applications have been focused on either operational or analytical workloads. We ended this chapter by describing how the performance of systems like SanssouciDB can be measured.

Chapter 8
Scaling SanssouciDB in the Cloud

Abstract This chapter describes why we believe the technology used in SanssouciDB is a good fit for Cloud Computing. We begin by defining Cloud Computing and then go on to describe the different types of applications that are suited for the cloud. Section 8.3 takes a provider's point of view and describes the type of infrastructure required to offer cloud services. We also examine multi-tenancy, replication, choice of hardware, and energy efficiency. Finally, we offer our conclusions regarding how in-memory computing will play an ever-increasing role in Cloud Computing.

Cloud Computing is gaining more momentum. Subscribing to services over the Internet is proving to be efficient, both from an economic and computational point of view. The provider aggregates many customers and can share all aspects of their IT infrastructure, including hardware, software, staffing, and the data center itself among a large number of customers. From a customer's point of view, the benefit comes from converting the fixed cost of owning and maintaining on-premise infrastructure into the variable and generally considerably lower, cost of renting it on demand. In this chapter, we introduce the term Cloud Computing in more detail, taking both the customer and provider view into account. We also describe application types that may especially benefit from the new possibilities offered by Cloud Computing and explain the role that in-memory database technology plays in the context of Cloud Computing.

Until recently enterprise software systems (ERP, human resources, CRM) have primarily been on-premise and single-tenant systems operated in closed environments and managed by customers' own IT departments. While outsourcing and focusing on core competencies has been a major trend since the 1990s, enterprise software systems have not generally been considered as part of this exercise because of their mission criticality. Consequently, the TCO of an enterprise software system is not only made up of the license costs for the software, but also includes many other costs. These include Capital Expenditures (CAPEX) for the computer and networking hardware to run the software as well as the Operational Expenditures (OPEX)

H. Plattner and A. Zeier, *In-Memory Data Management*,
DOI: 10.1007/978-3-642-29575-1_8, © Springer-Verlag Berlin Heidelberg 2012

for maintenance and support personnel, energy, cooling, and so on of these systems (see Sect. 1.3).

The need to cut costs in order to remain competitive, in combination with technological progress, is changing this, and Software-as-a-Service (SaaS) is becoming a viable and therefore more widely employed business model for enterprise software.

As a concept, SaaS has come a long way from the time-sharing systems of the late 1960s, via the Application Service Provider (ASP) approaches at the beginning of the new millennium when software first left the premises of the users [183], to today's solutions that host software and infrastructure at the large data centers of service providers. Companies access such applications as a service via the Internet.

The TCO for the companies using SaaS can thus be reduced since they only have to pay for using the service—either subscription-based or on a pay-per-use basis—but not for their own server hardware or their own data center. Of course, the service provider's price will include these costs but because of economies of scale, the cost for an individual customer can be reduced when compared to the on-premise deployment model.

An important leverage point is the ability to integrate application development with that of its supporting infrastructure. As a result, it becomes possible to highly optimize one for the other and vice versa. The more resources that are shared among the service provider's customers, the higher the economies of scale will be with respect to software, hardware, energy and administration costs [90].

8.1 What Is Cloud Computing?

As Cloud Computing is relatively new, there is no commonly agreed-upon definition, yet. Recurring aspects in many definitions include the provisioning of services over the Internet and an infrastructure that is reliable and has the scalability to provide these services (see [66] or [46]).

Depending on the type of service that is provided (or the level on which virtualization takes place) by this infrastructure, we can distinguish among the following:

- *IaaS*: Infrastructure-as-a-Service clouds offer OS infrastructures as a service.
- *PaaS*: Platform-as-a-Service clouds offer development platforms as a service.
- *SaaS*: Software-as-a-Service clouds offer applications as a service.

In the following, we will focus on the Software-as-a-Service aspect: particularly, the clouds that are providing application services to their clients, or more precisely data-centric enterprise application services. For a potential customer, the subscription-based or usage-based charging of services primarily means a conversion of CAPEX into OPEX, along with a possible overall reduction of IT costs. Moving to the cloud does not necessarily lead to lower costs than buying the respective hardware and software and then operating it on-premise. Rather, the real economic benefit of Cloud Computing is transferring the risk of over-provisioning (under-utilization) and under-provisioning (saturation) that is present in the on-premise case, thereby removing the need for an up-front capital commitment and potential surplus costs due to wrong estimates. This becomes possible because of the elasticity feature in Cloud Computing: cloud resources can be added or removed with fine granularity (one server at a time, for instance) and within a very short time frame (normally within min) versus the long lead times of growing (and shrinking) an on-premise solution. To summarize, there are five aspects of Cloud Computing that, in combination, set this approach apart from other existing concepts (see also [182]):

- Availability of resources, on demand, to the cloud user in a seemingly unlimited amount.
- Adaptability of used resources to the actual needs of the cloud user.
- Payment only for resources that were actually used or subscribed to by the cloud user on a short-term basis.
- The cloud provider operates and controls the cloud as a central instance.
- Software upgrades and enhancements are easier to manage in a hosted environment.

Regarding the accessibility of clouds, private and public clouds can be distinguished. A cloud is public whenever it is available to the general public. The services of a public cloud will be available in a pay-per-use manner. In contrast, a private cloud only allows restricted access: for example, all data centers of a global company can be understood as a typical case for a private cloud. Sometimes, this distinction is not quite clear. Imagine a cloud hosting SaaS enterprise resource planning solutions available to any client willing to pay.

8.2 Types of Cloud Applications

As we have already pointed out, the principles of Cloud Computing can be applied at different levels. Since we concentrate on applications, it is of interest to us what types of such applications may especially profit from the possibilities and in turn be a special driver for Cloud Computing. In the literature several types of applications are discussed that may play such a role [64, 82]:

- *Web 2.0 applications*: such applications must be highly available to their users but also depend on large amounts of data that need to be combined depending on the specific social context (or local context if the application is a mobile one) in which the mobile user currently navigates. Cloud Computing offers the storage space required for the different data sources that can be scaled-out depending on the current requirements. This especially allows start-up companies to avoid the initial investment in a server infrastructure and to use their money for its core processes.

- *Business analytics*: for the most part, analytical applications operate on large amounts of data returning aggregates or interesting patterns for decision making. Consequently, they are both processor- and memory-intensive while running and can largely benefit from the elasticity and the accompanying cost advantages of pay-per-use offered by Cloud Computing.
- *Compute-intensive parallel applications*: applications that are comprised in large parts of parallelizable tasks can benefit from Cloud Computing. The tasks can then be executed in a short time span using the vast processing capacities of the cloud while at the same time keeping to a tight budget frame because of the cost-effectiveness of Cloud Computing's elasticity.

The question of whether an application should be put into the cloud needs to be answered in its specific context. Inhibiting factors can be the sensitivity of the data in combination with the level of security that can be guaranteed in the cloud (also in combination with privacy and legal concerns), the economic risk of provider lock-in: because of proprietary APIs the cloud user can only switch providers at significant cost, mostly due to moving data in and out the cloud. The cloud user also has the choice of selecting different levels of outsourcing that are reflected in the levels of virtualization. For SaaS, the cloud user does not own the application but merely rents it. Examples of this approach are on-demand enterprise solutions, such as SAP Business ByDesign [151]. With the other two levels (PaaS, IaaS), the cloud user still owns the application and the platform, respectively, and uses the corresponding infrastructure of a cloud provider to run it. Existing Cloud Computing infrastructures that allow these two latter forms include the following:

- *Google File System (PaaS)*: a scalable distributed file system for large distributed data-intensive applications. The system provides fault tolerance while running on inexpensive commodity hardware, and it delivers high aggregate performance to a large number of clients [59].
- *MapReduce (PaaS)*: an implementation for processing and generating large data sets. Users specify a map function that processes a key/value pair to generate a set of intermediate key/value pairs, and a reduce function that merges all intermediate values associated with the same intermediate key. Many real- world tasks are expressible in this model. Programs written in this functional style are automatically parallelized and executed on a large cluster of commodity machines. The run-time system takes care of the details of partitioning the input data, scheduling the program's execution across a set of machines, handling machine failures, and managing the required inter-machine communication. This allows programmers without any experience in parallel and distributed systems to utilize the resources of a large distributed system [42].
- *IBM Smart Business Services (PaaS)*: a cloud portfolio meant to help clients turn complex business processes into simple services. To accomplish this, Smart Business Services brings sophisticated automation technology and self-service to specific digital tasks as diverse as software development and testing, desktop and device management, and collaboration.

- *Amazon EC2 (IaaS)*: a web service interface that allows the creation of virtual machines, thereby obtaining and configuring capacity with minimal difficulty. Its users are provided with complete control of the rented computing resources of Amazon's computing environment [7].
- *GoGrid (IaaS)*: a cloud infrastructure service, hosting Linux and Windows virtual machines managed by a multi-server control panel.
- *FlexiScale (IaaS)*: the utility computing platform launched by XCalibre Communications in 2007. Users are able to create, start, and stop servers as they require rapid deployment where needed. Both Windows and Linux are supported on the FlexiScale platform.

8.3 Cloud Computing from the Provider Perspective

"Cloud is a deployment detail, not fundamental. Where you run your software and what software you run are two different decisions, and you need to make the right choice in both cases", says Michael Olson, CEO at Cloudera [189]. For those users who decide against hosting an application on their own premises, Cloud Computing represents a transfer of risk from the user to the cloud provider with respect to resource dimensioning. This also means that the costs that this transfer represents for the cloud provider has to be less than the price obtained for providing the respective cloud services to its customers (the cloud users) on the market. This can only be achieved by economies of scale of building and operating extremely large data centers based on statistical multiplexing and bulk purchasing of the respective hardware and/or software. When calculating the costs for providing cloud services, hardware-, software-, and customer-related costs must be considered. Hardware-related costs comprise, for example, the depreciation on the hardware itself: the energy consumption of the hardware, including its cooling, the cost for the operating system and its maintenance, as well as the network. Software-related costs comprise the costs of the application including the database, its setup, its maintenance, upgrading to new releases, backups and its optimal operation. Customer-related costs are, for example, the customization of the application, business configuration and application management. Besides these cost aspects, a cloud provider has to consider much more when realizing a cloud, namely achieving good sharing of resources between multiple customers (that is, multi-tenancy) or providing an API for its adaptive data center.

8.3.1 Multi-Tenancy

Multi-tenancy is a technical concept that is frequently adopted in cloud computing with the goal of lowering total cost of ownership (TCO) on the end of the cloud service provider. In other words, multi-tenancy is a means to lower the cost of operating a cloud service, thus increasing the margin of the product. The term itself is is a metaphor which uses an anology from the world of real estate: the idea is that the tenant of an apartment is given the illusion if *isolation*, that is the illusion that the he or she can make full use of the apartment. When the tenant is not at home, however, the landlord may choose to temporarily repurpose the apartment.

So the term multi-tenancy refers to the consolidation of several customers onto the same server farm. With a single-tenant system, each customer company will have its own applications and its own databases on server machines dedicated to them. Costs for the service provider in such cases will be relatively high, as the promised economies of scale cannot be fully realized. In contrast, on multi-tenant systems different customers share the same resources of an operational system on the same machine. Multi-tenancy is an optimization for hosted services in environments with multiple customers (that is, tenants) [91]. Salesforce.com first employed this technique on a large scale.

Multi-tenancy can be used on all layers of cloud computing: an IaaS provider will typically use virtualization to increase the utilization of its servers. While this seems to be a simple example for multi-tenancy at the first glance, such multi-tenancy is hard to implement since the *workload* put on the virtual machines of the different tenants is completely heterogeneous. In other words, the cloud service provider does not know what a tenant is using its virtual machine for. Therefore, elaborate monitoring must be in place to ensure that the system as a whole has enough resources to serve all tenants that are assigned to it. Due to the heterogenety of the tenants, the cloud providers ability to detect load spikes and react to them accordingly (and thus continue providing the perfect illusion of isolation) will only be so good.

A PaaS provider faces a similar challenge since it has no or little control over the applications that customers run on their platform (which, again, are potentially heterogeneous in nature). This challenge is only slightly alleviated by the fact that all applications run on top of the same platform.

True homogeneous multi-tenancy can only be found in SaaS applications, where all tenants run not only on the same platform but also run the same app. The difference among multiple tenants is solely in the workload that the tenants exert on the application.

In effect, multi-tenancy improves utilization by resource pooling and thereby eliminates the need for over-provisioning to meet the maximal load requirements of each of the service provider's customers. By providing a single administration framework for the whole system, multi-tenancy can improve the efficiency of system management [91].

To live up to their full potential, multi-tenant systems comprise capabilities to scale both up and out. Scaling up is the consolidation of several customers onto the operational system of the same server machine. Scaling out, on the other hand, refers to having the administration framework span a number of such machines or a data center. Scale-out is thus of great importance for the success of the SaaS business model as the capacity of a single machine cannot be enhanced beyond a certain limit without running into unjustifiable hardware costs. This also means that tenant migration among multiple machines in a data center must be possible [91]. Multi-tenancy almost always occurs at the database layer of the provided service, especially because of the stateless nature of application servers for highly scalable Web applications [9].

8.3.1.1 How Much Consolidation Is Possible?

To get a sense for how much consolidation can be achieved with multi-tenancy in the database, let us consider the following example: one fact table in the warehouse data we got from one of the larger SAP customers holds all the sales records for three business years. The table has approximately 360 million records. Once imported into SanssouciDB, the complete size of the table in main memory is less than 3 GB. While this particular customer is a Fortune 500 company, a typical SaaS customer would be factors smaller, since the SaaS model targets small and medium enterprises. At the same time, blade server boards that allow up to 2 TB of physical memory are commercially available at the time of this writing. Obviously, the potential for consolidation is huge.

Note that data set size is not the only factor restraining main memory. For query processing, it is often necessary to materialize intermediate results in main memory. As the number of concurrent users on a server increases, more memory must be reserved for intermediate results. A model describing the relation between data set size and number of concurrent users across all tenants on a machine is introduced in [156].

8.3.1.2 Requirements for Multi-Tenancy

As multi-tenancy is an important cornerstone to achieve the economies of scale that makes the SaaS business model successful, several requirements must be fulfilled by multi-tenant systems or more precisely—because multi-tenancy is primarily a database layer issue—by multi-tenant databases; especially as provisioning and operating databases normally is a cost-intensive task. There should be an administration infrastructure that is able to maintain metadata about customers (for example, contact/location information, location information in the server farm, activated features) and allow querying this data in a uniform query language. Base schemas for the provided services specifying all their default fields should be maintained and the customer-specific extensions should be kept as part of the metadata. In addition, public information, such as ZIP codes or current exchange rates, should be made available to all customers as well as the administration framework. The latter could potentially be outside of the database and should offer support for tenant migration, that is, for moving the data of one customer from one server machine to another within the server farm, as described in [156].

Another important issue for a service provider is keeping the software up to date. Here, rolling upgrades have to be possible. This means that not all servers are updated to a new application version at the same time but one after another. This is essential

for gaining experience with the new application version under realistic workloads be
fore putting it into global productive use.

8.3.1.3 Implementing Multi-Tenancy

When implementing multi-tenancy in the database three distinct approaches can be
used: shared machines, shared database instance, and shared tables. With increasing
granularity of sharing, the ability to pool resources and to execute bulk administrative
operations improves. On the other hand, with each step, isolation among customers
is reduced which may have a negative impact on security and on fairness of resource
sharing. In the following, we will describe the alternatives in more detail [91]:

Shared Machine

In this implementation scheme each customer has its own database process and these
processes are executed on the same machine: that is, several customers share the same
server.

One advantage of this approach is the very good isolation among tenants. A major
limitation is that this approach does not support memory pooling and each database
needs its own connection pool: communication sockets cannot be shared among
databases. Economies of scale are further limited by the infeasibility of executing
administrative operations in bulk because of the greater isolation among the different
database processes. On the other hand, this isolation makes customer migration from
one machine to another a straightforward that merely entails moving files between
machines.

Shared Database Instance

In the shared database instance approach each customer has its own tables and sharing
takes place on the level of the database instances. Here, connection pools can be
shared between customers and pooling of memory is considerably better than with
the previous approach. Bulk execution of administrative operations across all tenants
within the database instance is possible. On the other hand, the isolation between
the customers is reduced. As a compromise, this approach seems to be the most
promising for implementing multi-tenant systems, especially in cases whe re one
base schema is used for all customers.

Shared Table

In this last approach sharing takes place on the level of database tables: data from
different customers is stored in one table, in order to identify to which customer data

belongs, a tenant identifier needs to be added as an additional attribute to each table. Customer-specific extensions of the base scheme can be achieved by adding a set of generic columns to each table.

The shared table approach performs best with regard to resource pooling. Scaling up takes place by adding new rows and is consequently only bound by the database's row number limitation. Of course, sharing of connection pools between customers is also possible. Administrative operations can be carried out in bulk by running queries over the column containing the tenant identifier.

Isolation is weak with this approach. As files contain mixed data from different customers, migration of a tenant is not easy. This intermingling also leads to contention between customers' queries and impacts the system's ability to realize query optimization. Security can only be ensured if assignment of access privileges is possible at the level of rows.

The techniques for dynamically scheduling queries for execution based on the current load situation described in Sect. 5.9 are directly applicable when using this approach.

As a historical note it is worth mentioning that SAP first adopted a shared table approach in SAP R/2, which was designed in 1979. In R/2, the requirement arose to separate business entities in large companies from each other, for example subcompanies in holding structures. This was realized by giving every sub-entity a dedicated client number.

8.3.1.4 Schema Evolution

As stated above, one requirement for multi-tenant databases is to cater the ability of individual tenants to make custom changes to the database tables. Note that these customers would in fact make changes to the application, which would then result in changes to the database tables. Here, the focus is on extensibility in the database layer.

In row-oriented databases, it is generally hard to maintain a large number of tables. As an example, IBM DB2 V9.1 allocates 4 kB of memory for each table so 100,000 tables consume 400 MB of memory up-front. Buffer pool pages are allocated on a per-table basis so there is great competition for the remaining cache space. As a result, the performance on a blade server begins to degrade beyond about 50,000 tables. Multi-tenancy is mostly realized using the shared table approach discussed above. Having all tenants share the same database tables complicates private tenant extensions. A detailed evaluation of schema evolution techniques on top of row-oriented databases can be found in [9].

In contrast, comparatively little management data must be stored per table when using a column store, since there is no paging and buffer pool infrastructure. In our setting, it is reasonable to give tenants their private tables. Tenant extensions and the accompanying data dictionary changes are therefore restricted to the tenants' individual scope and do not interfere with other tenants. Row-oriented databases usually do not allow data definition operations while the database is online [9].

Note that this is much easier for in-memory column databases and thus happens while the database is online: the table is locked only for the amount of time that is required to complete the data definition operation. It should be noted that this data definition operation is restricted entirely to the metadata of the table. The document vector and dictionary of a column are not created until the first write occurs. This late materialization is also a performance optimization, since many operations in a column store require full column scans. When scanning a column that has been created but not yet populated, the scan operator can simply skip the column. From an operational perspective, lazy materialization of columns significantly speeds up administrative operations such as creating a new tenant (see Sect. 5.7).

While metadata for extensions is managed on a per-tenant level, the metadata for common tables (without/before applying extensions) is stored in shared memory. The metadata for all the base tables is read from shared memory. Tenant-local changes (as they happen in the case of extensions) are stored in the tenants' private metadata spaces.

8.3.2 Low-End Versus High-End Hardware

To realize the data center that will provide the cloud services two different hardware approaches can be distinguished: the low-end versus the high-end hardware approach. In the low-end hardware approach, a very large number (up to millions and more) of relatively inexpensive computers, which are specifically built from commodity components, are used, while the high-end hardware approach only employs high-performance, high-availability server machines/blades in a much smaller number (up to the ten thousand range), but provide similar capacity because of their higher individual processing power. It is still under debate [102], which of the two approaches would generally be better suited for Cloud Computing, but the answer probably lies with the application domain.

For mixed workload enterprise applications of the kind that we describe in this book, a cloud-variant of SanssouciDB would be the system of choice. Since SanssouciDB massively leverages parallelism by exploiting multi-core processors, it becomes clear that an application cloud for hosting mixed workload enterprise applications cannot be based on low-end hardware. However, the possibility of leveraging economies of scale is not restricted to using cheap hardware. Not only do larger servers provide a higher potential for consolidation; they also offer more headroom for accomodating occasional load spikes of individual tenants. Deciding how much headroom should be reserved for load spikes and how much of a servers capacity should be allocated to tenants is a difficult optimization problem, which is also one focus of our current research work [156].

Note that the question of what type of hardware the service provider should select is orthogonal to whether multi-tenancy will be employed, and, if so, on what level. For the case of a shared machine multi-tenant architecture we would like to refer the reader to Sect. 4.5, where we have provided with a detailed discussion on the consequences of running a main memory database in a virtual environment on either a commodity system with an FSB architecture or a high-end system with a dedicated on-chip memory controller per socket.

For non-enterprise applications the hardware decisions of a cloud service provider might look different: if the user can tolerate losing an end-user session without far-reaching consequences, such as losing a query for a search engine, the low-end approach is viable. If losing the same session has repercussions on data integrity and/or a larger amount of work of one or even more users (for example, restarting a transaction and losing all data entries), the high-end approach with its much higher availability is most certainly preferable.

8.3.3 Replication

Another heavily debated topic in recent work on Cloud Computing is replication for scalability and high-availability as well as the question of how to keep multiple copies synchronized (and at what cost should this be done [51]). Numerous systems have evolved to alleviate the scalability limits of relational databases for cloud environments, for example PNUTS [28], Bigtable [23], or Dynamo [43], for all of which disk is still the limiting factor on replica consistency.

In-memory architectures are ideal for running data-centric enterprise applications in the cloud. In fact, replication across two nodes where the data is kept in main memory both offers higher throughput and is more predictable than writing to disk on a single node.

The math behind this statement is simple: While high-end disks today have a latency of $13,000\,\mu s$ before one can even start writing, writing to main memory on two nodes costs two times $10\,ns$ plus the cost for the data transmission over the network ($500\,\mu s$) plus a couple of mutex lock/unlock operations ($0.1\,\mu s$ each) to get to write to the socket, thus less than $600\,\mu s$ in total. There might be contention around the network links (similarly to contention for disk I/O) but they can be scaled horizontally fairly easy. Also, research is currently underway with the goal of achieving round trip times of less than $10\,\mu s$ for remote procedure calls [129].

Particularly for scan-intensive applications such as data warehousing, for which one typically uses column-oriented databases, in-memory architectures seem a practicable approach given a maximum theoretical memory bandwidth of $25.6\,GB$ per second per socket on current Intel platforms. In contrast to disk-based column stores, in-memory column databases do not require a-priori knowledge about the queries and, more importantly, do not need to physically store and maintain multiple potentially overlapping collections of columns (projections) to avoid CPU stalls during scans due to disk I/O [165].

Workloads with relatively short read and write transactions are also a good candidate for in-memory architecture of a distributed database. The idea is to have a deterministic serial ordering of transactions. In a distributed setting deterministic serial transaction ordering reduces the communication overhead, which is traditionally necessary for deadlock detection and two-phase commit and thus enables better scalability. The experiments in [171] show that it is in-memory technology that makes the deterministic serial ordering viable: a deterministic serialization strategy produces more locking overhead than a non-deterministic strategy. The overhead grows with average transaction execution time. With in-memory technology average transaction execution time shrinks dramatically and as a consequence the locking overhead produced by the deterministic strategy is quite small. It is small enough to gain performance and stability by reducing the communication overhead.

8.3.4 Energy Efficiency by Employing In-Memory Technology

Energy consumption is another important factor for data center operations. For example, in [137] the trade-off between total power consumption and performance for a TPC-H warehousing benchmark is discussed by considering multiple storage options for the database tables with different cost characteristics. The evaluated configurations range across multiple parallel disk setups (between 6 and 100 hard disks, SSD, and main memory). The TPC-H data has been generated with a scale factor of 64; hence the fact table has a size of 64 GB. A server with 16 GB of main memory (which can be used as a cache for all configurations) is used to run an Oracle 11g database. Next to the database server is a storage system that holds the data. The benchmark has been carried out with 6, 16, 32, and 100 parallel disks in the storage server. For the two other configurations, the database server was enhanced with two SSDs with a capacity of 320 GB each and 64 GB of main memory (instead of 16 GB as for the other configurations), respectively. Note that 64 GB are not enough to hold the entire data set in memory since the operating system and the database process itself need to allocate a certain fraction of the memory.

Figure 8.1 shows the results of the benchmark. The vertical axis shows the amount of kilowatt-hours consumed for running all the queries defined in TPC-H sequentially.

The following conclusions can be drawn from the chart:

1. The in-memory configuration offers the best performance and consumes the least amount of power among the tested configurations.
2. Only the configuration with 100 parallel disks provides a throughput near the throughput observed on the main memory variant. The 100 disk variant consumes more than three times the power than the main memory variant.
3. A storage server with 100 parallel disks and the I/O capabilities specified in [137] costs approximately US $ 100,000 to date, whereas 64 GB of main memory retail at US $ 4,200.

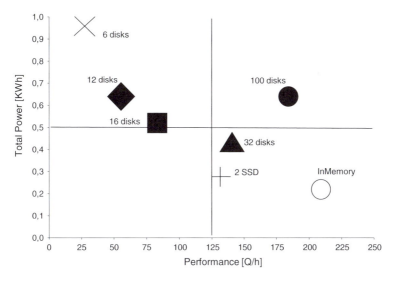

Fig. 8.1 Power consumption versus throughput [13]

8.4 Conclusion

This chapter offered a broad overview of Cloud Computing and considered the perspectives of both the customer and the cloud service provider. We provided a classification of types of applications that can be found in cloud environments and discussed the challenges to service providers of running a data center, from selecting the right hardware to hosting issues such as virtualization. We discussed these aspects from the point of view of in-memory computing, which we believe has an important role to play in the field of Cloud Computing. Cloud Computing is widely perceived as technology for processing long-lived jobs on large data sets using commodity hardware, for example, log file analyses at Facebook [172]. It is, however, equally attractive when many relatively small data sets need to be handled: as is the case of such applications as the class of Enterprise SaaS solutions, where the service provider develops an application and operates the system that hosts it.

Chapter 9
The In-Memory Revolution has Begun

Abstract This final chapter demonstrates that our vision of an in-memory database for large-scale enterprise applications is already in the process of becoming reality. The first product that implements many of the concepts of SanssouciDB is the new in-memory data management solution released by SAP at the end 2010. Companies can begin using in-memory applications today and benefit from a massive performance increase. We show how this transition can occur seamlessly and without disruption. The chapter closes with the first customer proof points by presenting application examples.

In the history of enterprise software, new technology has generally only been deployed to new products or the latest release of an existing product. Radical changes to how data is organized and processed usually involve major adaptations throughout the entire stack of enterprise applications. This does not align well with the evolutionary modification schemes of business-critical customer systems. Consequently, the dispersion and adoption of technological advantages in existing on-premise system landscapes is often delayed by long-term customer IT roadmaps and conservative upgrade policies. However, considering the significant advantages that in-memory data management can bring to enterprises, we think that an immediate adoption of this technology is critical for the existing customer base. Instead of merely moving forward, as is commonly done in the high-tech industry, we must also look backwards and cater our solutions to existing systems that are already out there. In the following, we demonstrate how in-memory technology can be implemented in existing enterprise systems, immediately and without disruption.

9.1 Risk-Free Transition to In-Memory Data Management

In a design thinking workshop conducted with IT students from the Hasso Plattner Institute in Potsdam in February 2010, we interviewed executives and IT professionals from different industries about how in-memory technology could change

H. Plattner and A. Zeier, *In-Memory Data Management*, 233
DOI: 10.1007/978-3-642-29575-1_9, © Springer-Verlag Berlin Heidelberg 2012

their daily business. The demand for an adoption plan that bypasses the regular release cycles became evident. While the increase in speed and performance would bring instant competitive advantage, companies expressed reservations about moving immediately to the next software release. Even in cases where resource issues provided no obstacle, participants shied away from being an early adopter or changing a stable system: "Let somebody else try it. If it is really proven, come back to us." was a commonly heard response.

Motivated by the results of this workshop, we proposed a transition process that allows customers to benefit from in-memory technology without changing their running system. The migration process we present is non-disruptive and fundamentally risk-free. Step by step, we seamlessly transform traditionally separated operational and analytical systems into what we believe is the future for enterprise applications: transactional and analytical workload handled by a single, in-memory database.

9.1.1 Operating In-Memory and Traditional Systems Side by Side

The first step in establishing the bypass is to install and connect the IMDB to the traditional DBMS of an existing operational system. An initial load then creates a complete and exact logical image of the existing business objects inside the IMDB, with the difference that the replicated data is stored in columns rather than in rows. First experiments with massively parallel bulk loads of customer data have shown that even for the largest companies this one-time initialization can be done in only a few hours.

Once the initial load has completed, the database loader of the ERP system then permanently manages the two storage systems in parallel: anything that happens in the traditional database is mirrored in the in-memory database. A parallel setup with two identical databases may, at first, appear to be a waste of resources, but the high compression rates in a column store considerably decrease the amount of main memory that is actually required (see Sect. 4.3). For 1 TB of customer data we found that 100 GB of in-memory storage capacity is sufficient. Providing the necessary main memory is affordable and yields instant benefits: from the first day on, traditional ETL processes can now run from the in-memory database through the ERP system and feed the traditional BI system. Optionally, the in-memory database can be placed underneath the existing BI system to recreate the traditional data warehouse in main memory and to achieve another gain in reporting performance (Fig. 9.1). Without having changed the customer software, we can speed up the generation of reports by orders of magnitude, and do this in an instant and absolutely risk-free manner.

Once the parallel setup is in place, the next step in the transition is to bypass the materialized views in the traditional analytical system and to connect the in-memory engine to the new real-time analytical BI applications this technology makes possible. Instead of taking the data out of the operational system through ETL, and reporting

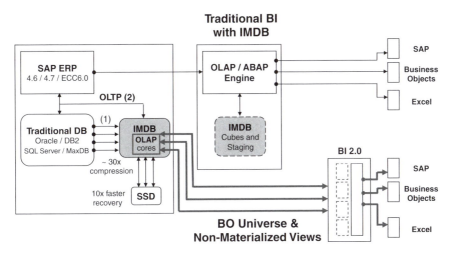

Fig. 9.1 Operating in-memory and traditional database systems side by side

against data in a BI system, we can start to use more intuitive user interaction tools and directly access the operational data from the new in-memory database shown on the left of Fig. 9.1. The advantages are obvious: the data cubes in a traditional BI system are usually updated on a weekly basis or even less frequently. However, executives often demand up-to-the-moment information on arbitrary combinations of data. With in-memory applications, we can deliver any of this information in seconds.

9.1.2 System Consolidation and Extensibility

So far, the transition process has not imposed any binding changes on the existing configuration. No extra load is put on the traditional database. The performance and functioning of the production system is not jeopardized. With the parallel installation of an in-memory data store underneath the traditional OLTP and OLAP systems, an immediate speed-up in reporting can be achieved at absolutely no risk.

As more analytical tasks run directly against the IMDB, the load on the new system components increases. Once customers are comfortable, the traditional OLAP components and the traditional OLTP database can be switched off. What remains is a clean, consolidated in-memory enterprise system that establishes a single source of truth for real-time, any-style business analytics. New applications can now be added as required to extend the functionality of the system on-premise or out of the cloud. Extensions to the data model are also simple to implement and can happen without any disruption. Adding new tables and new attributes to existing tables in the column store is done on the fly. This will speed up release changes dramatically (Fig. 9.2).

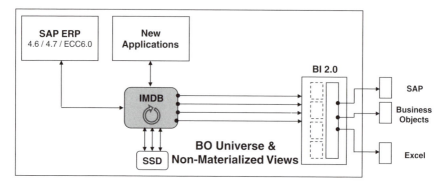

Fig. 9.2 Consolidated system architecture with SAP's new in-memory database

9.2 Customer Proof Points

With SanssouciDB, we have presented a concept that we believe is the ideal in-memory data management engine for the next generation of real-time enterprise applications. Its technological components have been prototypically implemented and tested individually in our research. While we expect to soon see the first systems that combine all the characteristics of SanssouciDB, companies can already leverage in-memory technology to improve the performance of their analytical systems. Optimizing the planning, prediction, and processes in enterprises with in-memory technology is possible today. In this section, we give examples of how businesses have already begun to benefit from real-time analytics enabled by in-memory technology. The following ramp-up use cases give a striking account of the feasibility, desirability, and viability of these applications.

In an early validation phase, SAP has loaded 275 million sales records of a global consumer goods company into an in-memory database. This enabled the analysis of the data along any dimensions at the speed of thought. Simple ad-hoc queries took no longer than 0.04 s. In general, a factor 100 increase in performance was achieved [109]. This reduction of response times to almost zero will have a significant impact on the way businesses are run, but just as important is the fact that this gain in performance is affordable. As discussed in the previous chapters, the cost of acquiring additional main memory storage is compensated by the reduced efforts in maintaining the complex traditional reporting systems. In the end, customers benefit from an overall decrease in TCO.

In-memory technology has also been integrated into SAP BusinessObjects Explorer, a BI solution to run product profitability checks, inventory planning, and other analytical processes in real-time. The on-demand solution SAP Business ByDesign also features in-memory technology, and more products are entering the market as this book is being written.

SAP's latest solution for analytical in-memory data management called HANA (High-Performance Analytical Appliance) uses many of the concepts we have described in this book, and is currently being rolled out to pilot customers in different industries. HANA's performance impact on common reporting scenarios is stunning. For example, the query execution times on 33 million customer records of a large financial service provider dropped from 45 min on the traditional DBMS to 5 s on HANA. The increase in speed fundamentally changes the company's opportunities for customer relationship management, promotion planning, and cross selling. Where once the traditional data warehouse infrastructure has been set up and managed by the IT department to pre-aggregate customer data on a monthly basis, HANA now enables end users to run live reports directly against the operational data and to receive the results in seconds.

In a similar use case, a large vendor in the construction industry is using HANA to analyze its nine million customer records and to create contact listings for specific regions, sales organizations, and branches. Customer contact listing is currently an IT process that may take two to three days to complete. A request must be sent to the IT department who must plan a background job that may take 30 min and the results must be returned to the requestor. With HANA, sales people can directly query the live system and create customer listings in any format they wish, in less than 10 s.

A global producer of consumer goods was facing limitations with its current analytical system in that it was not able to have brand and customer drilldowns in the same report. This rigidity was solved with HANA. Not only have the query execution times for profitability analysis dropped from 10 min to less than 10 s, with the new user interface, based on SAP BusinessObjects Explorer, it is now also possible to do arbitrary drilldowns, select flexible time periods, and to export, customize, and update the reports in other applications (for example, Microsoft Excel).

The next two subsections present two customer case studies with the Charité—Universitätsmedizin Berlin and Hilti in Liechtenstein. We have collaborated with both companies for many years and they are among the first to leverage the advantages of in-memory computing.

9.2.1 Charité—Universitätsmedizin Berlin

Charité—Universitätsmedizin Berlin is the largest university hospital in Europe. It is located in Germany's capital Berlin and is split across three locations with a total area of more than 607,000 sqm. As one of the largest employers in the capital region, Charité has 14,500 employees, which includes 3,750 scientific and medical staff, 4,225 nursing staff, and 250 professors. In addition, more than 7,500 students are matriculated. With more than 133,000 inpatient cases and more than 564,000 outpatient cases per year, Charité generates 1.3 billion EUR annual turnover.

9.2.1.1 Objective of Using HANA

Charité's objectives are to improve treatment times of patients while reducing costs for operations. A diversified IT landscape across various institutions, departments and vendor systems in combination with the steadily increasing amount of clinical data makes it hard to identify relevant information and to have it at hand. Medical doctors require instant access to all relevant information for decision making when interacting with the patient directly. In addition, researchers require flexible ways to supervise cohorts of patients and to identify candidates for participation in certain clinical studies.

9.2.1.2 Explain the HANA Solution

Hana Oncolyzer is a proof of concept to support medical doctors and researchers of the oncology department at Charité.

Medical doctors have an instant search engine available on their iPad device. With the help of fuzzy search and synonym tables the tool supports instant search on various data sources containing unstructured and structured data. The transparent combination of search criteria helps to derive knowledge that was gathered in the patient data. A dedicated patient screen combining all data makes personalized medicine possible. The doctor's decision making is supported by seamlessly integrating third-party data sources when searching for certain terms in the patient's data, for example, the drugs a patient is taking. As a result, the doctor has all relevant information at hand, making their decision-making process more reliable. Researchers at Charité can make qualitative statements about participating in a clinical study anywhere at any time. The iPad prototype offers flexible filters that can be applied to the real data set of patient data. This helps to assess whether patients with certain criteria are available and whether participating in a clinical study is possible. Without the prototype the search has to be performed with an existing BI system, which requires the IT department to prepare a concrete report and to extract, transform and load the relevant data from the various data sources. The prototype reduced the lookup time from two days per patient down to minutes.

9.2.1.3 Summary

The Charité customer story shows how in-memory computing is used to support diverse tasks. Instead of traditional enterprise applications, the focus lies on patient treatment and therapy evaluation, analysis of patient cohorts, and extraction of knowledge from structured and unstructured data sources. Due to the requirements to have all patient data at hand, real-time analysis of patient data supports the daily life of researchers and medical doctors in the oncology department.

9.2.2 Hilti

Hilti is one of the world's leading companies in providing products, systems and services to construction professionals. Founded 1941 in Liechtenstein, Hilti is present in more than 120 countries with around 20.000 employees. Hilti adopts a direct sales model and around 60% of all Hilti employees work in the sales force; sales representatives, field engineers and customer service/call center employees who create more than 220000 customer contacts every day. Hilti's annual revenue was approximately 4 billion CHF in 2010. Within Hilti, global IT is fully aligned with the global business process organization. Hilti runs SAP globally, based on a single instance ERP solution and globally harmonized processes and data covering 95% of all business transactions worldwide. Due to the direct sales model the reporting capabilities—especially for the sales force—play an important role within Hilti's IT business. Hilti's Business Warehouse solution manages around 10TB of data, 13000 active users, 250000 report calls per month—of which 70000 are pre-calculated and 180000 are dialog-based—and is based on a global standardized set of KPIs and reports.

9.2.2.1 Objective of Using HANA

Hilti's objectives for using HANA are twofold: Increase the productivity of its internal IT organization, and deliver value for the business organization. With HANA, Hilti expects to significantly simplify its reporting architecture and reduce its TCO. As an example, Hilti currently maintains more than 100 data cubes for its profitability analysis for performance reasons. With HANA, the redundant storage of aggregated data becomes largely obsolete, simplifying the architecture and increasing efficiency by freeing up resources in warehouse maintenance that have been largely occupied with readjusting aggregates and data marts or introducing new KPIs. Besides increasing the productivity of its internal IT, Hilti expects to realize benefits for the business organization by optimizing business processes. With reports now running orders of magnitude faster, Hilti can run analytics in real-time, thereby changing their way business is done in the future. As an example, providing reports in real-time to smart-devices will enable Hilti's sales reps to check open order statuses, product availability or customers credit information during customer interactions, and thus further improve customer experience. Furthermore, the increased performance with HANA will allow Hilti to simulate the effects of product-mix or price changes on the margin to enable better-informed and more flexible decision-making.

9.2.2.2 Explain the HANA Solution

Two of the first IT-processes at Hilti that have benefited from in-memory data management with HANA are profitability analysis and fleet management. Profitability analysis is a tool to analyze revenues and costs on the level of products and sales units. It provides information on the short-term profitability of the organization and thus supports management in its decision-making. The results from profitability analysis are fed into the business warehouse for further reporting. The original solution operates on roughly 450 Mio. records; new records are loaded three times per day. To achieve an acceptable performance, data is structured in more than 100 cubes, split by organization and year, with up to six cubes for each organization for performance reasons. Particularly at each month end, Hilti faces a high volume of new records and very long loading times. With HANA, Hilti can avoid aggregating data in cubes, which significantly reduces load latency time and data redundancy. Furthermore, the maintenance effort of the HANA-based solution is very low and no special maintenance is required at year-end. First test queries on the HANA-based solution for profitability analysis have shown outstanding results. From 320 s in the original solution, the run time has been reduced to just about a second.

Besides profitability analysis, Hilti also applies in-memory data management with HANA to its fleet management solution. Instead of simply buying a Hilti product, customers can get Hilti equipment for a fixed monthly charge that covers all tool, service and repair costs. All tools in the fleet are replaced at regular intervals with tools of the latest generation. This process is managed by Hilti's fleet management system, which operates on 45 Mio. data records. In the original solution, new records could not be loaded daily due to technical restrictions. Hence, Hilti was only able to load new records once a week on weekends, taking more than 4 h each time and Hilti customer service reps could not get an up-to-date view on which tools a customer had collected or returned to Hilti. With the HANA-based solution, Hilti could integrate new records into their reporting at any time and provide a real-time report about which customers have which tools to its 7.500 sales representatives. Time-consuming, manual investigations to get a consistent view on the tools held by each customer become largely obsolete. In summary, the advantages of the HANA-based solution are up-to-date data in reports, no necessity to store redundant data in aggregates, almost no load latency, and dramatically reduced maintenance effort which offers the possibility to reduce headcount and thus, has a direct bottom-line impact.

9.2.2.3 Summary

Based on the early experiences with HANA, Hilti's mid-term objective is to consolidate transactional and analytical data stores as much as possible, so that data optimized for reporting does not have to be stored in a separate database, but can be used to calculate the operations in real time. To reach this objective Hilti envisions running its ERP, CRM and warehouse solution on a single instance in-memory database. This way, Hilti expects to further improve the efficiency of its internal IT organization and allow its business organization to profit from better decision-making support by providing real-time information anywhere and at any time, and independent of the employee's access device.

9.3 Conclusion

When we started our research in this field six years ago, we were often confronted with great skepticism. The performance increase achieved through in-memory technology is often difficult to grasp and convey. Even after reading this book, the full impact of in-memory enterprise data management might not be evident for everyone to see. Understanding all the technical details, concepts, and implications is a good starting point, but in order to fully understand the consequences of such an approach, it is imperative that one personally experiences how the interaction with in-memory changes enterprise applications. The ability to get the answer to any query within seconds or less, to ask any kind of business-relevant question, and all this on the most up-to-the-moment data will change the enterprises of today and future. We are experiencing the beginning of a revolution—in-memory technology truly marks an inflection point for enterprise applications. Real-time enterprise computing is finally here.

About the Authors

Prof. Dr. h.c. Hasso Plattner is a co-founder of SAP AG, where he served as the CEO until 2003 and has since been chairman of the supervisory board. SAP AG is today the leading provider of enterprise software solutions. In his role as chief software advisor, he concentrates on defining the mid- and long-term technology strategy and direction of SAP. Hasso Plattner received his diploma in communications engineering from the University of Karlsruhe. In recent years, he has been focusing on teaching and research in the field of business computing and software engineering at large. In 1998, he founded the Hasso Plattner Institute (HPI) in Potsdam, Germany. At the HPI, approximately 480 students are currently pursuing their Bachelors' and Masters' degrees in IT Systems Engineering with the help of roughly 50 professors and lecturers. The HPI currently has about 100 PhD candidates. Hasso Plattner leads one of the research groups at HPI which focuses mainly on In-Memory Data Management for Enterprise Applications and Human-Centered Software Design and Engineering (see epic.hpi.uni-potsdam.de).

Dr. Alexander Zeier graduated from the University of Wuerzburg in business management and successfully completed his studies in information technology at the TU Chemnitz. He worked for a few years as a strategic IT consultant, before gaining his Ph.D. in Supply Chain Management (SCM) at the University of Erlangen-Nuremberg. He has 20 years experience with IT/SAP Systems and started working for SAP in 2002 as product manager with overall responsibility for the SCM Software, SAP's first large In-Memory Application. Since 2006 he has been Deputy Chair Enterprise Platform and Integration Concepts of Prof. Hasso Plattner at the Hasso Plattner Institute in Potsdam, focusing on real-time In-Memory Enterprise Systems. During that time he has also been Executive Director for the European Section of the MIT Forum for Supply Chain Innovation. Since March 2012 Dr. Zeier has been working at the Massachusetts Institute of Technology (MIT) as Visiting Professor, lecturing and conducting research in the area of In-Memory Technology & Applications, and Supply Chain Innovation. He is the author of more than 150 journal articles and papers and has also published six books on IT and SAP (see zeier.mit.edu).

H. Plattner and A. Zeier, *In-Memory Data Management*,
DOI: 10.1007/978-3-642-29575-1, © Springer-Verlag Berlin Heidelberg 2012

Glossary

ACID Property of a database management system to always ensure atomicity consistency, isolation, and durability of its transactions.

Active Data Data of a business transaction that is not yet completed and is therefore always kept in main memory to ensure low latency access.

Aggregation Operation on data that creates a summarized result for example, a sum, maximum, average, and so on. Aggregation operations are common in enterprise applications.

Analytical Processing Method to enable or support business decisions by giving fast and intuitive access to large amounts of enterprise data.

Application Programming Interface (API) Aninterface for application programmers to access the functionality of a software system.

Atomicity Database concept that demands that all actions of a transaction are executed or none of them.

Attribute A characteristic of an entity describing a certain detail of it.

Availability Characteristic of a system to continuously operate according to its specification measured by the ratio between the accumulated time of correct operation and the overall interval.

Available-to-Promise (ATP) Determining whether sufficient quantities of a requested product will be available in current and planned inventory levels at a required date in order to allow decision making about accepting orders for this product.

Batch Processing Method of carrying out a larger number of operations without manual intervention.

Benchmark A set of operations run on specified data in order to evaluate the performance of a system.

H. Plattner and A. Zeier, *In-Memory Data Management*,
DOI: 10.1007/978-3-642-29575-1, © Springer-Verlag Berlin Heidelberg 2012

Blade Server in a modular design to increase the density of available computing power.

Business Intelligence Methods and processes using enterprise data for analytical and planning purposes or to create reports required by management.

Business Logic Representation of the actual business tasks of the problem domain in a software system.

Business Object Representation of a real-life entity in the data model for example, a purchasing order.

Cache A fast but rather small memory that serves as buffer for larger but slower memory.

Cache Coherence State of consistency between the versions of data stored in the local caches of a CPU cache.

Cache-Conscious Algorithm An algorithm is cache conscious if program variables that are dependent on hardware configuration parameters (for example cache size and cache-line length) need to be tuned to minimize the number of cache misses.

Cache Line Smallest unit of memory that can be transferred between main memory and the processor's cache. It is of a fixed size which depends on the respective processor type.

Cache Miss Afailed request for data from a cache because it did not contain the requested data.

Cache-Oblivious Algorithm An algorithm is cache oblivious if no program variables that are dependent on hardware configuration parameters (for example cache size and cache-line length) need to be tuned to minimize the number of cache misses.

Characteristic-Oriented Database System A database system that is tailored towards the characteristics of special application areas. Examples are text mining stream processing and data warehousing.

Cloud Computing An IT provisioning model which emphasizes the ondemand, elastic pay-per-use rendering of services or provisioning of resources over a network.

Column Store Database storage engine that stores each column (attribute) of a table sequentially in a contiguous area of memory.

Compression Encoding information in such a way that its representation consumes less space in memory.

Concurrency Control Techniques that allow the simultaneous and independent execution of transactions in a database system without creating states of unwanted incorrectness.

Consistency Database concept that demands that only correct database states are visible to the user despite the execution of transactions.

Consolidation Placing the data of several customers on one server machine database or table in a multi-tenant setup.

Cube Specialized OLAP data structure that allows multi-dimensional analysis of data.

Customer Relationship Management (CRM) Business processes and respective technology used by a company to organize its interaction with its customers.

Data Aging The changeover from active data to passive data.

Data Center Facility housing servers and associated ICT components.

Data Dictionary Metadata repository.

Data Layout The structure in which data is organized in the database that is the database's physical schema.

Data Mart A database that maintains copies of data from a specific business area for example, sales or production, for analytical processing purposes.

Data Warehouse A database that maintains copies of data from operational databases for analytical processing purposes.

Database Management System (DBMS) A set of administrative programs used to create maintain and manage a database.

Database Schema Formal description of the logical structure of a database.

Demand Planning Estimating future sales by combining several sources of information.

Design Thinking A methodology that combines an end-user focus with multidisciplinary collaboration and iterative improvement. It aims at creating desirable user-friendly, and economically viable design solutions and innovative products and services.

Desirability Design thinking term expressing the practicability of a system from a human-usability point of view.

Dictionary In the context of this book the compressed and sorted repository holding all distinct data values referenced by SanssouciDB's main store.

Dictionary Encoding Light-weight compression technique that encodes variable length values by smaller fixed-length encoded values using a mapping dictionary.

Differential Buffer A write-optimized buffer to increase write performance of the SanssouciDB column store. Sometimes also referred to as differential store or delta store.

Distributed System A system consisting of a number of autonomous computers that communicate over a computer network.

Dunning The process of scanning through open invoices and identifying overdue ones in order to take appropriate steps according to the dunning level.

Durability Database concept that demands that all changes made by a transaction become permanent after this transaction has been committed.

Enterprise Application A software system that helps an organization to run its business. A key feature of an enterprise application is its ability to integrate and process up-to-the-minute data from different business areas providing a holistic real-time view of the entire enterprise.

Enterprise Resource Planning (ERP) Enterprise software to support the resource planning processes of an entire company.

Extract-Transform-Load (ETL) Process A process that extracts data required for analytical processing from various sources then transforms it (into an appropriate format, removing duplicates, sorting, aggregating, etc.) such that it can be finally loaded into the target analytical system.

Fault Tolerance Quality of a system to maintain operation according to its specification even if failures occur.

Feasibility Design thinking term expressing the practicability of a system from a technical point of view.

Front Side Bus (FSB) Bus that connects the processor with main memory (and the rest of the computer).

Horizontal Partitioning The splitting of tables with many rows into several partitions each having fewer rows.

Hybrid Store Database that allows mixing column- and row-wise storage.

In-Memory Database Adatabase system that always keeps its primary data completely in main memory.

Index Data structure in a database used to optimize read operations.

Insert-Only New and changed tuples are always appended already existing changed and deleted tuples are then marked as invalid.

Inter-Operator Parallelism Parallel execution of independent plan operators of one or multiple query plans.

Intra-Operator Parallelism Parallel execution of a single plan operation independently of any other operation of the query plan.

Isolation Database concept demanding that any two concurrently executed transactions have the illusion that they are executed alone. The effect of such an isolated execution must not differ from executing the respective transactions one after the other.

Join Database operation that is logically the cross product of two or more tables followed by a selection.

Latency The time that a storage device needs between receiving the request for a piece of data and transmitting it.

Locking A method to achieve isolation by regulating the access to a shared resource.

Logging Process of persisting change information to non-volatile storage.

Main Memory Physical memory that can be directly accessed by the central processing unit (CPU).

Main Store Read-optimized and compressed data tables of SanssouciDB that are completely stored in main memory and on which no direct inserts are allowed.

MapReduce A programming model and software framework for developing applications that allows for parallel processing of vast amounts of data on a large number of servers.

Materialized View Result set of a complex query which is persisted in the database and updated automatically.

Memory Hierarchy The hierarchy of data storage technologies characterized by increasing response time but decreasing cost.

Merge Process Process in SanssouciDB that periodically moves data from the write-optimized differential store into the main store.

Metadata Data specifying the structure of tuples in database tables (and other objects) and relationships among them in terms of physical storage.

Mixed Workload Database workload consisting both of transactional and analytical queries.

Multi-Core Processor A microprocessor that comprises more than one core (processor) in a single integrated circuit.

Multi-Tenancy The consolidation of several customers onto the operational system of the same server machine.

Multithreading Concurrently executing several threads on the same processor core.

Network Partitioning Fault Fault that separates a network into two or more subnetworks that cannot reach each other anymore.

Node Partial structure of a business object.

Normalization Designing the structure of the tables of a database in such a way that anomalies cannot occur and data integrity is maintained.

Object Data Guide A database operator and index structure introduced to allow queries on whole business objects.

Online Analytical Processing (OLAP) see Analytical Processing.

Online Transaction Processing (OLTP) see Transactional Processing.

Operational Data Store Database used to integrate data from multiple operational sources and to then update data marts and/or data warehouses.

Padding Approach to modify memory structures so that they exhibit better memory access behavior but requiring the trade-off of having additional memory consumption.

Passive Data Data of a business transaction that is closed/completed and will not be changed anymore. For SanssouciDB it may therefore be moved to non-volatile storage.

Prefetching A technique that asynchronously loads additional cache lines from main memory into the CPU cache to hide memory latency.

Query Request sent to a DBMS in order to retrieve data manipulate data, execute an operation, or change the database structure.

Query Plan The set and order of individual database operations derived by the query optimizer of the DBMS, to answer an SQL query.

Radio Frequency Identification (RFID) Wireless technology to support fast tracking and tracing of goods. The latter are equipped with tags containing a unique identifier that can be readout by reader devices.

Real Time In the context of this book defined as, within the timeliness constraints of the speed-of-thought concept.

Real-Time Analytics Analytics that have all information at its disposal the moment they are called for (within the timeliness constraints of the speed of thought concept).

Recoverability Quality of a DBMS to allow for recovery after a failure has occurred.

Recovery Process of re-attaining a correct database state and operation according to the database's specification after a failure has occurred.

Relational Database A database that organizes its data in relations (tables) as sets of tuples (rows) having the same attributes (columns) according to the relational model.

Response Time at the Speed of Thought Response time of a system that is perceived as instantaneous by a human user because of his/her own mental processes. It normally lies between 550 ms and 750 ms.

Return on Investment (ROI) Economic measure to evaluate the effciency of an investment.

Row Store Database storage engine that stores all tuples sequentially that is each memory block may contain several tuples.

Sales Analysis Process that provides an overview of historical sales numbers.

Sales Order Processing Process with the main purpose of capturing sales orders.

SanssouciDB The in-memory database described in this book.

Scalability Desired characteristic of a system to yield an efficient increase in service capacity by adding resources.

Scale-out Capable of handling increasing workloads by adding new machines and using these multiple machines to provide the given service.

Scale-up Capable of handling increasing workloads by adding new resources to a given machine to provide the given service.

Scan Database operation evaluating a simple predicate on a column.

Scheduling Process of ordering the execution of all queries (and query plan operators) of the current workload in order to maintain a given optimality criterion.

Sequential Reading Reading a given memory block by block.

Shared Database Instance Multi-tenancy implementation scheme in which each customer has its own tables and sharing takes place on the level of the database instances.

Shared Machine Multi-tenancy implementation scheme in which each customer has its own database process and these processes are executed on the same machine that is several customers share the same server.

Shared Table Multi-tenancy implementation scheme in which sharing takes place on the level of database tables that is data from different customers is stored in one and the same table.

Shared Disk All processors share one view to the non-volatile memory but computation is handled individually and privately by each computing instance.

Shared Memory All processors share direct access to a global main memory and a number of disks.

Shared Nothing Each processor has its own memory and disk(s) and acts independently of the other processors in the system.

Single Instruction Multiple Data (SIMD) A multiprocessor instruction that applies the same instructions to many data streams.

Smart Grid An electricity network that can intelligently integrate the behavior and actions of all users connected to it—generators consumers and those that do both in order to efficiently deliver sustainable, economic and secure electricity supplies.

Software-as-a-Service (SaaS) Provisioning of applications as cloud services over the Internet.

Solid-State Drive (SSD) Data storage device that uses microchips for nonvolatile high-speed storage of data and exposes itself via standard communication protocols.

Speed-Up Measure for scalability defined as the ratio between the time consumed by a sequential system and the time consumed by a parallel system to carry out the same task.

Star Schema Simplest form of a data warehouse schema with one fact table (containing the data of interest for example, sales numbers) and several accompanying dimension tables (containing the specific references to view the data of interest, for example, state, country, month) forming a star-like structure.

Stored Procedure Procedural programs that can be written in SQL or PL/SQL and that are stored and accessible within the DBMS.

Streaming SIMD Extensions (SSE) An Intel SIMD instruction set extension for the x86 processor architecture.

Structured Data Data that is described by a data model for example, business data in a relational database.

Structured Query Language (SQL) A standardized declarative language for defining querying, and manipulating data.

Supply Chain Management (SCM) Business processes and respective technology to manage the flow of inventory and goods along a company's supply chain.

Table A set of tuples having the same attributes.

Tenant (1) A set of tables or data belonging to one customer in a multitenant setup. (2) An organization with several users querying a set of tables belonging to this organization in a multi-tenant setup.

Thread Smallest schedulable unit of execution of an operating system.

Three-tier Architecture Architecture of a software system that is separated in a presentation a business logic, and a data layer (tier).

Time Travel Query Query returning only those tuples of a table that were valid at the specified point in time.

Total Cost of Ownership (TCO) Accounting technique that tries to estimate the overall life-time costs of acquiring and operating equipment for example, software or hardware assets.

Transaction A set of actions on a database executed as a single unit according to the ACID concept.

Transactional Processing Method to process every-day business operations as ACID transactions such that the database remains in a consistent state.

Translation Lookaside Buffer (TLB) A cache that is part of a CPU's memory management unit and is employed for faster virtual address translation.

Trigger A set of actions that are executed within a database when a certain event occurs for example a specific modification takes place.

Tuple A real-world entity's representation as a set of attributes stored as element in a relation. In other words a row in a table.

Unstructured Data Data without data model or that a computer program cannot easily use (in the sense of understanding its content). Examples are word processing documents or electronic mail.

Vertical Partitioning The splitting of the attribute set of a database table and distributing it across two (or more) tables.

Viability Design thinking term expressing the practicability of a system from an economic point of view.

View Virtual table in a relational database whose content is defined by a stored query.

Virtual Machine A program mimicking an entire computer by acting like a physical machine.

Virtualization Method to introduce a layer of abstraction in order to provide a common access to a set of diverse physical and thereby virtualized resources.

References

1. Abadi, D.: Query Execution in Column-Oriented Database Systems. Ph.D. thesis, MIT (2008)
2. Abadi, D., Boncz, P., Harizopoulos, S.: Column oriented database systems. PVLDB **2**, 1664–1665 (2009)
3. Advanced Micro Devices, I.: HyperTransport Technology I/O Link, a High-Bandwidth I/O Architecture. URL whitepaper. Advanced Micro Devices Inc., Sunnyvale, CA (2001)
4. Agrawal, R., Ailamaki, A., Bernstein, P.A., Brewer, E.A., Carey, M.J., Chaudhuri, S., Doan, A., Florescu, D., Franklin, M.J., Garcia-Molina, H., Gehrke, J., Gruenwald, L., Haas, L.M., Halevy, A.Y., Hellerstein, J.M., Ioannidis, Y.E., Korth, H.F., Kossmann, D., Madden, S., Magoulas, R., Ooi, B.C., O'Reilly, T., Ramakrishnan, R., Sarawagi, S., Stonebraker, M., Szalay, A.S., Weikum, G.: The claremont report on database research. SIGMOD Rec. **37**, 9–19 (2008). doi: http://doi.acm.org/10.1145/1462571.1462573. URL http://doi.acm.org/10.1145/.1462573
5. Ailamaki, A., DeWitt, D., Hill, M.: Data page layouts for relational databases on deep memory hierarchies. VLDB J. **11**, 198–215 (2002)
6. Alsberg, P.: Space and time savings through large data base compression and dynamic restructuring. Proc. IEEE **102**, 1114–1122 (1975)
7. Amazon: Amazon Elastic Compute Cloud. http://aws.amazon.com/ec2/ (2010). Retrieved 14 Jan 2011
8. Amdahl, G.: Validity of the single processor approach to achieving large scale computing capabilities. AFIPS Conf. Proc. **30** 483–485 (1967)
9. Aulbach, S., Grust, T., Jacobs, D., Kemper, A., Rittinger, J.: Multi-tenant databases for software as a service: schema-mapping techniques. In: SIGMOD, pp. 1195–1206 (2008)
10. Banerjee, J., Kim,W., Kim, H., Korth, H.F.: Semantics and implementation of schema evolution in object-oriented databases. In: SIGMOD, pp. 311–322 (1987)
11. Barham, P., Dragovic, B., Fraser, K., Hand, S., Harris, T.L., Ho, A., Neugebauer, R., Pratt, I.,Warfield, A.: Xen and the art of virtualization. In: SOSP, pp. 164–177 (2003)
12. Becker, S.A.: Developing Quality Complex Database Systems: Practices, Techniques, and Technologies. Idea Group Inc., Hershey (2001)
13. Behling, S., Bell, R., Farrell, P., Holthoff, H., O'Connell, F., Weir, W.: The POWER4 Processor Introduction and Tuning Guide. URL IBM Redbooks (2001)
14. Binnig, C., Faerber, F., Hildenbrand, S.: Dictionary-based order-preserving string compression for main-memory column stores. In: SIGMOD, pp. 283–296 (2009)
15. Bog, A., Krueger, J., Schafner, J.: A composite benchmark for online transaction processing and operational reporting. In: AMIGE (2008)
16. Bog, A., Plattner, H., Zeier, A.: A mixed transaction processing and operational reporting benchmark. Inf. Syst. Frontiers J. **13** 301–304 (2010)

17. Borr, A.: Robustness to Crash in a Distributed Database: A Non Shared-Memory Multi-Processor Approach (1984)
18. Botezatu, B.: The Future of Processors, Painted in Multi-Core Colors. URL http://news.softpedia.com/news/The-Future-of-Processors-Painted-in-Multi-Core-Colors-78143.shtml (2008). Retrieved 14 Jan 2011
19. Brooks, F.: The Mythical Man-Month. Addison-Wesley, Boston (1975)
20. Ceri, S., Negri, M., Pelagatti, G.: Horizontal data partitioning in database design. In: SIGMOD, pp. 128–136 (1982)
21. Chamberlin, D., Boyce, R.: Sequel: A structured English query language. In: SIGFIDET (now SIGMOD), pp. 249–264 (1974)
22. Chamoni, P., Gluchowski, P.: Analytische Informationssysteme: Business Intelligence-Technologien und Anwendungen. Springer, Heidelberg (2006)
23. Chang, F., Dean, J., Ghemawat, S., Hsieh,W.C.,Wallach, D.A., Burrows, M., Chandra, T., Fikes, A., Gruber, R.E.: Bigtable:A distributed storage system for structured data. ACM Trans. Comput. Syst. **26**, 1–4 (2008)
24. Chang, S.: Database decomposition in a hierarchical computer system. In: SIGMOD, pp. 48–53 (1975)
25. Chen, Z., Gehrke, J., Korn, F.: Query optimization in compressed database systems. In: SIGMOD, pp. 271–282 (2001)
26. Cieslewicz, J., Ross, K.: Adaptive aggregation on chip multiprocessors. In: VLDB, pp. 339–350 (2007)
27. Codd, E.: A relational model of data for large shared data banks. Commun. ACM **13**, 377–387 (1970)
28. Cooper, B. F., Ramakrishnan, R., Srivastava, U., Silberstein, A., Bohannon, P., Jacobsen, H., et al.: PNUTS: Yahoo!'s hosted data serving platform. PVLDB **1**, 1277–1288 (2008)
29. Corporation, S.P.E.: SPEC CPU2006. URL Benchmark website: http://www.spec.org/cpu2006/ (2006). Retrieved 14 Jan 2011
30. Corporation, S.P.E.: SPECviewperf 11. URL Benchmark website: http://www.spec.org/gwpg/gpc.static/vp11info.html (2006). Retrieved 14 Jan 2011
31. Corporation, S.P.E.: SPECmail2009. URL Benchmark website: http://www.spec.org/mail2009/ (2009). Retrieved 14 Jan 2011
32. Corporation, S.P.E.: SPECweb2009. URL Benchmark website: http://www.spec.org/web2009/ (2009). Retrieved 14 Jan 2011
33. Corporation, S.P.E.: SPECjEnterprise2010. URL http://www.spec.org/jEnterprise2010/ (2010). Retrieved 14 Jan 2011
34. Corporation, S.P.E.: SPECvirt. URL Benchmark website: http://www.spec.org/virt_sc2010/ (2010). Retrieved 14 Jan 2011
35. Corporation, W.D.: Specification for the Seraial AT 6 Gb/s VelociRaptor Enterprise Hard Drives. URL Product website: http://wdc.custhelp.com/cgibin/wdc.cfg/php/enduser/std_adp.php?p_faqid=5377&p_created= (2010). Retrieved 14 Jan 2011
36. Council, T.P.P.: URL http://www.tpc.org (2010). Retrieved 14 Jan 2011
37. Cross, R.L.: ITJ Foreword Q1, 2002, Intel Hyper-Threading Technology (2002)
38. Date, C.: An Introduction to Database Systems, 6th edn. Addison-Wesley, Boston (1995)
39. David, J., Schuff, D., St Louis, R.: Managing your IT Total Cost of Ownership. Commun. ACM **45**, 101–106 (2002)
40. Dean, J.: Designs, Lessons and Advice from Building Large Distributed Systems. URL http://www.slideshare.net/xlight/google-designs-lessons-and-advicefrom-building-large-distributed-systems (2009). Retrieved 14 Jan 2011
41. Dean, J.: Large-scale distributed systems at Google: current systems and future directions. In: LADIS (2009)
42. Dean, J., Ghemawat, S.: MapReduce: simplified data processing on large clusters. In: OSDI, pp. 137–150 (2004)

43. DeCandia, G., Hastorun, D., Jampani, M., Kakulapati, G., Lakshman, A., Pilchin, A., Sivasubramanian, S., Vosshall, P., Vogels, W.: Dynamo: Amazon's highly available key-value store. In: SIGOPS, pp. 205–220 (2007)
44. Devices, A.M.: AMD. URL http://www.amd.com. Retrieved 14 Jan 2011
45. DeWitt, D., Gray, J.: Parallel database systems: the future of high performance database systems. Commun. ACM **35** 85–98 (1992)
46. Dikaiakos, M.D., Katsaros, D., Mehra, P., Pallis, G., Vakali, A.: Cloud computing: distributed Internet computing for IT and scientific research. IEEE Internet Comput. **13**, 10–13 (2009)
47. Eager, D., Zahorjan, J., Lozowska, E.: Speedup versus efficiency in parallel systems. IEEE Trans. Comput. **38**, 408–423 (1989)
48. Elmasri, R., Navathe, S.: Fundamentals of Database Systems, 5th edn. Addison-Wesley Longman Publishing, Boston (2006)
49. Exasol: EXASolution Highspeed Database. URL Product website, http://www.exasol.com/ (2010). Retrieved 14 Jan 2011
51. Fine, C.: Clockspeed :Winning Industry Control in the Age of Temporary Advantage. Basic Books, New York (1998)
51. Florescu, D., Kossmann, D.: Rethinking cost and performance of database systems. SIGMOD Record **38**, 43–48 (2009)
52. Flynn, M.: Very high-speed computing systems. Proc. IEEE **54**(12), 1901–1909 (1966)
53. French, C.: Teaching an OLTP database Kernel advanced data warehousing techniques. In: ICDE, pp. 194–198 (1997)
54. Fujitsu: Speicher-Performance Xeon 5500 (Nehalem EP) basierter PRIMERGY Server (2009). URL Whitepaper, version 1.0
55. Fujitsu: Speicher-Performance Xeon 7500 (Nehalem EX) basierter Systeme (2010). URL Whitepaper, Version 1.0
56. Garcia-Molina, H., Salem, K.: Main memory database systems: an overview. TKDE **4**, 509–516 (1992)
57. Garcia-Molina, H., Ullman, J., Widom, J.: Database Systems: The Complete Book. Prentice Hall Press, Upper Saddle River (2008)
58. Gates, B.: Information At Your Fingertips. URL Keynote Address, Fall/-COMDEX, Las Vegas, Nevada (1994)
59. Ghemawat, S., Gobioff, H., Leung, S.: The Google file system. In: SOSP, pp. 29–43 (2003)
60. Graefe, G.: Query evaluation techniques for large databases. ACM Comput. Surv. **25**, 73–170 (1993)
61. Gray, J.: A Transaction Model. In: Colloquium on Automata, Languages and Programming, pp. 282–298 (1980)
62. Gray, J.: The Benchmark Handbook for Database and Transaction Processing Systems. Morgan Kaufmann Publishers, San Mateo (1993)
63. Gray, J.: Tape is Dead, Disk is Tape, Flash is Disk, RAM Locality is King. URL http://www.signallake.com/innovation/Flash_is_Good.pdf (2006). Retrieved 14 Jan 2011
64. Gray, J.: Distributed Computing Economics. ACM Queue **6**, 63–68 (2008)
65. Gray, J., Reuter, A.: Transaction Processing: Concepts and Techniques. Morgan Kaufmann, San Fransisco (1993)
66. Grossman, R. L. (2009). The case for cloud computing. IT Prof. **11**, 23–27
67. Grund, M., Krueger, J., Plattner, H., Zeier, A., Madden, S., Cudre-Mauroux, P.: HYRISE— A hybrid main memory storage engine. In: VLDB (2011)
68. Grund, M., Krueger, J., Tinnefeld, C., Zeier, A.: Vertical partition for insert-only scenarios in enterprise applications. In: IEEM (2009)
69. Grund, M., Schaffner, J., Krueger, J., Brunnert, J., Zeier, A.: The effects of virtualization on main memory systems. In: DaMoN, pp. 41–46 (2010)
70. Gschwind, M.: The Cell Broadband Engine: Exploiting Multiple Levels of Parallelism in a Chip Multiprocessor (2006)

71. Habich, D., Boehm, M., Thiele, M., Schlegel, B., Fischer, U., Voigt, H., Lehner, W.: Next generation database programming and execution environment. In: VLDB (2011)
72. Haerder, T., Rahm, E.: Datenbanksysteme: Konzepte und Techniken der Implementierung, 2 Auflage. Springer, Berlin (2001)
73. Haerder, T., Reuter, A.: Principles of transaction-oriented database recovery. ACM Comput. Surv. **15**, 287–317 (1983)
74. Hagel, J., Brown, J. S., Davison, L.: The Power of Pull: How Small Moves, Smartly Made, Can Set Big Things in Motion.Basic Books, New York (2010)
75. Hamilton, J.: Cost of Power in Large-Scale Data Centers. URL http://perspectives. mvdirona.com/2008/11/28/CostOfPowerInLargeScaleDataCenters.aspx (2008). Retrieved 14 Jan 2011
76. Hankins, R.A., Patel, J.M.: Data morphing: an adaptive, cache-conscious storage technique. In: VLDB, pp. 417–428 (2003)
77. Hare, C.: PC Hardware Links. URL Web page, http://mysite.verizon.net/pchardwarelinks/ main.htm (2010). Retrieved 26 July 2010
78. Hellerstein, J.: Datalog Redux: Experience and Conjecture. In: PODS, pp. 1–2 (2010)
79. Hellerstein, J.M., Stonebraker, M., Hamilton, J.: Architecture of a Database System. Now Publishers Inc., Hanover (2007)
80. Hennessy, J.L., Patterson, D.A.: Computer Architecture. Morgan Kaufmann, Boston (2006)
81. Hitachi: Hitachi Global Storage Technologies, Ultrastar 15K450. URL Product website, http://www.hitachigst.com/internal-drives/enterprise/ultrastar/ultrastar-15k450 (2010). Retrieved 14 Jan 2011
82. Hohpe, G., Woolf, B.: Enterprise Integration Patterns: Designing, Building, and Deploying Messaging Solutions. Addison-Wesley Longman Publishing Co. Inc., Boston (2003)
83. Inmon, B.: Data mart does not equal dataWarehouse. DMReview May 1998. URL DM Rev. Mag. (1998)
84. Inmon, B.: ODS Types. Information management magazine January 2000. URL Inf. Manag. Mag. (2000)
85. Inmon, B.: Operational and informational reporting. Information management magazine July 2000. URL Inf. Manag. Mag. (2000)
86. Intel: An Introduction to the Intel QuickPath Interconnect. URL http://www. intel.com/technology/quickpath/introduction.pdf (2009). Retrieved 14 Jan 2011
87. Intel: Intel R X25-E SATA Solid State Drive. URL http://download.intel.com/design/flash/ nand/extreme/319984.pdf (2009). Retrieved 14 Jan 2011
88. Intel: Microprocessor Quick Reference Guide. URL http://www.intel.com/pressroom/ kits/quickrefyr.htm (2010). Retrieved 14 Jan 2011
89. Intel: Moore's Law: Made Real by Intel Innovation. URL Web site: http://www. intel.com/technology/mooreslaw/ (2010). Retrieved 14 Jan 2011
90. Jacobs, D.: Enterprise software as service. ACM Queue **3**, 36–42 (2005)
91. Jacobs, D., Aulbach, S.: Ruminations on multi-tenant databases. In: BTW, pp. 514–521 (2007)
92. Kimball, R.: Surrogate Keys—Keep Control Over Record Identifiers by Generating New Keys for the Data Warehouse. URL http://www.rkimball.com/html/articles_search/ articles1998/9805d05.html (1998). Retrieved 14 Jan 2011
93. Kimball, R., Caserta, J.: The DataWarehouse ETL Toolkit. Wiley, New York (2004)
94. Knolmayer, G., Mertens, P., Zeier, A., Dickersbach, J.: Supply Chain Management Based on SAP Systems: Architecture and Planning Processes (2009)
95. Knuth, D.: The Art of Computer Programming, vol. 3, 2nd edn., Sorting and Searching. Addison Wesley Longman Publishing Co. Inc., Boston (1998)
96. Kossow, R.: TCO-Wirtschaftlichkeir von IT-Systemen. URL http://www.erpmanager.de/ magazin/artikel_ 1339_tco_total_cost_ownership_wirtschaftlichkeit.html (2007). Retrieved 14 Jan 2011
97. Krueger, J., Grund, M., Tinnefeld, C., Zeier, A., Plattner, H.: Optimizing write performance for read optimized databases. In: DASFAA (2010)

98. Krueger, J., Grund, M.,Wust, J., Zeier, A., Plattner, H.: Merging differential updates in in-memory column store. In: DBKDA (2011)
99. Krueger, J., Grund, M., Zeier, A., Plattner, H.: Enterprise application-specific data management. In: EDOC, pp. 131–140 (2010)
100. Krueger, J., Tinnefeld, C., Grund, M., Zeier, A., Plattner, H.: A case for online mixed workload processing. In: DBTest (2010)
101. Laming, D.: Information Theory of Choice-Reaction Times. Academic Press, New York (1968)
102. Lang, W., Patel, J., Shankar, S.: Wimpy node clusters: what about non-Wimpy workloads? In: DaMoN, pp. 47–55 (2010)
103. Lee, J., Kim, K., Cha, S.: Differential logging: a commutative and associative logging scheme for highly parallel main memory databases. In: ICDE, pp. 173–182 (2001)
104. Lee, S., Moon, B.: Design of flash-based DBMS: an in-page logging approach. In: SIGMOD, pp. 55–66 (2007)
105. Lemke, C., Sattler, K.U., Faerber, F.: Compression techniques for column-oriented BI accelerator solutions. In: BTW, pp. 468–497 (2009)
106. Lemke, C., Sattler, K.U., Faerber, F., Zeier, A.: Speeding up queries in column stores—a case for compression data. In: DaWaK, pp. 117–129 (2010)
107. Lerner, B.S., Habermann, A.N.: Beyond schema evolution to database reorganization. In: OOPSLA/ECOOP, pp. 67–76 (1990)
108. Ma, H.: Distribution Design for Complex Value Databases (2007)
109. Magura, S.: Warten war frueher. SAP Spectrum 2, 8–9 (2010)
110. Manegold, S., Boncz, P.A., Kersten, M.L.: Generic database cost models for hierarchical memory systems. In: VLDB, pp. 191–202 (2002)
111. Mehldorn, K., Sanders, P.: Algorithms and Data Structures. Springer, Berlin (2008)
112. Mertens, P.: Integrierte Informationsverarbeitung. Gabler, Wiesbaden (2009)
113. Mertens, P., Zeier, A.: ATP—available-to-promise. Wirtschaftsinformatik 41, 378–379 (1999)
114. Microsoft: Microsoft PowerPivot. URL Product website: http://www.powerpivot.com (2010). Retrieved 14 Jan 2011
115. Microsoft: Scalability. URL http://msdn.microsoft.com/enus/library/aa29217228v=VS.7129.aspx (2010). Retrieved 14 Jan 2011
116. Molka, D., Hackenberg, D., Schone, R., Muller, M.: Memory performance and cache coherency effects on an Intel Nehalem multiprocessor system. In: PACT, pp. 261–270 (2009)
117. Moore, G.: Cramming more components onto integrated circuits. Electron. Mag. 38, 114–117 (1965)
118. Moura, E., Ziviani, N., Baeza-Yates, R., Navarro, G.: Fast and flexible word searching on compressed text. ACM TOIS 18, 113–139 (2000)
119. Moutsos, K.: IMS at 40: Stronger than ever. IBM Database Mag. 4 (2008)
120. MP, S.: High Performance Computing Virtualization, Virtual SMP. URL Product website, http://www.scalemp.com/ (2010). Retrieved 14 Jan 2011
121. Naffziger, S., Warnock, J., Knapp, H.: SE2 when processors hit the power wall (or When the CPU Hits the Fan). In: ISSCC (2005)
122. Navathe, S.B., Ceri, S., Wiederhold, G., Dou, J.: Vertical partitioning algorithms for database design. ACM Trans. Database Syst. 9, 680–710 (1984)
123. North, K.: Terabytes to Petabytes: Reflections on 1999–2009. URL http://www.drdobbs.com/blog/archives/2010/01/terabytes_to_pe.html (2010). Retrieved 14 Jan 2011
124. O'Neil, P., Winter, R., French, C., Crowley, D., McKenna, W.: Data warehousing lessons from experience. In: ICDE, p. 294 (1998)
125. O'Neil, P.E., O'Neil, E.J., Chen, X.: The Star Schema Benchmark (SSB). URL http://www.cs.umb.edu/poneil/StarSchemaB.pdf (2007). Retrieved 14 Jan 2011
126. O'Neil, P.E., O'Neil, E.J., Chen, X., Revilak, S.: The star schema benchmark and augmented fact table indexing. In: TPCTC, pp. 237–252 (2009)
127. Oracle: Oracle E-Business Suite Standard Benchmark. URL http://www.oracle.com/apps-benchmark/results-166922.html (2010). Retrieved 14 Jan 2011

128. Orlowski, A.: POWER4 Debuts in IMB Regatta: Big Blue's Big Bang Eschews SMP Numbers Game. URL The Register, http://www.theregister.co.uk/2001/10/04/power4_debuts_in_ibm_regatta/ (2001), Retrieved 14 Jan 2011
129. Ousterhout, J. K., Agrawal, P., Erickson, D., Kozyrakis, C., Leverich, J., Mazières, D., et al.: The case for RAMClouds: scalable high-performance storage entirely in DRAM. Oper. Syst. Rev. **43**, 92–105 (2009)
130. Pacioli, F.L.: Treatise on Double-Entry Bookkeeping. Translated by Pietro Crivell (1494). Institute of Book-Keepers, London (1924)
131. Panchenko, O., Karstens, J., Plattner, H., Zeier, A.: Precise and Scalable Querying of Syntactical Source Code Patterns Using Sample Code Snippets and a Database. In: Proceedings of the International Conference on Program Comprehension, pp. 41–50. IEEE Computer Society, Los Alamitos, CA, USA (2011)
132. Panchenko, O., Plattner, H., Zeier, A.: Mapping terms in application and implementation domains. In: Proceedings of the Workshop on Software Reengineering (2011)
133. Paraccel: Paraccel—Column-Oriented DBMS for Decision Support and Complex Processing. URL Product website: www.paraccel.com/ (2010). Retrieved 14 Jan 2011
134. Pedersen, T.B., Jensen, C.S.: Multidimensional database technology. Computer **34**, 40–46 (2001)
135. Plattner, H.: A common database approach for OLTP and OLAP using an in-memory column database. In: SIGMOD, pp. 1–2 (2009)
136. Plattner, H.: SanssouciDB: an in-memory database for mixed-workload processing. In: BTW (2011)
137. Poess, M., Nambiar, R.O.: Tuning servers, storage and database for energy efficient data warehouses. In: ICDE, pp. 1006–1017 (2010)
138. Poess, M., Smith, B., Kollar, L., Larson, P.: TPC-DS, taking decision support benchmarking to the next level. In: SIGMOD, pp. 582–587 (2002)
139. Ponniah, P.: Data Warehousing Fundamentals: A Comprehensive Guide for IT Professionals, vol. 1. Wiley, New York (2001)
140. Power, D.: A Brief History of Decision Support Systems. URL http://dssresources.com/history/dsshistory.html (2010). Retrieved 14 Jan 2011
141. QlikView: Business Intelligence (BI) Software Solutions—Business Intelligence Reporting Software. URL Product website: http://www.qlikview.com/ (2010). Retrieved 14 Jan 2011
142. Raden, N.: Exploring the Business Imperative of Real-Time Analytics. URL Teradata Whitepaper (2003)
143. Raden, N.: Business Intelligence 2.0: Simpler, More Accessible, Inevitable. URL Intelligent enterprise, http://intelligent-enterprise.com (2007). Retrived 14 Jan 2011
144. Rahm, E.: Mehrrechner-Datenbanksysteme—Grundlagen der Verteilten und Parallelen Datenbankverarbeitung. Addison-Wesley, Boston (1994)
145. Rao, J., Ross, K.: Making B+- trees cache conscious in main memory. SIGMOD Rec. **29**, 475–486 (2000)
146. Roddick, J.: A Survey of Schema Versioning Issues for Database Systems. Information and Software Technology **37**, 383–393 (1995). URL citeseer. ist.psu.edu/roddick95survey.html
147. Rossberg, J., Redler, R.: Pro Scalable .NET 2.0 Application Designs. Designing .NET 2.0 Enterprise Applications from Conception to Deployment. Apress, New York (2005)
148. Roth, M., Van Horn, S.: Database compression. SIGMOD Rec. **22**, 31–39 (1993)
149. Salomon, D.: Data Compression. Springer, New York (2006)
150. Sanders, P., Transier, F.: Intersection in integer inverted indices. In: ALENEX (2007)
151. SAP: SAP Business By Design. URL http://www.sap.com/germany/sme/solutions/businessmanagement/businessbydesign/index.epx. Retrieved 14 Jan 2011
152. SAP: SAP BusinessObjects Explorer, Explore Your Business at the Speed of Thought. URL Product website: http://www.sap.com/solutions/sapbusinessobjects/large/businessintelligence/search-navigation/explorer/index.epx. Retrieved 14 Jan 2011
153. SAP: SAP Netweaver BusinessWarehouse Accelerator. URL Product Website, http://www.sdn.sap.com/irj/sdn/bwa (2008). Retrieved 14 Jan 2011

154. SAP: SAP Standard Application Benchmark. URL http://www.sap.com/solutions/benchmark/index.epx (2010). Retrieved 14Jan 2011

155. Schaffner, J., Bog, A., Krueger, J., Zeier, A.: A hybrid row-column OLTP database architecture for operational reporting. In: BIRTE (Informal Proceedings) (2008)

156. Schaffner, J., Eckart, B., Jacobs, D., Schwarz, C., Plattner, H., Zeier, A.: Predicting in-memory database performance for automating cluster management tasks. In: ICDE (2011)

157. Schaffner, J., Eckart, B., Schwarz, C., Brunnert, J., Jacobs, D., Zeier, A., Plattner, H.: Simulating multi-tenant olap database clusters. In: Datenbanksysteme in Business, Technologie und Web (BTW 2011), 14. Fachtagung des GI-Fachbereichs Datenbanken und Informationssysteme (DBIS), Proceedings, Kaiserslautern, Germany (2011)

158. Schapranow, M.P., Kuehne, R., Zeier, A.: Enabling real-time charging for smart grid infrastructures using in-memory databases. In: 1st IEEE LCN Workshop on Smart Grid Networking Infrastructure (2010)

159. Schapranow, M.P., Zeier, A., Plattner, H.: A dynamic mutual RFID authentication model preventing unauthorized third party access. In: The 4th International Conference on Network and System Security (2010)

160. Schapranow, M.P., Zeier, A., Plattner, H.: A formal model for enabling RFID in pharmaceutical supply chains. In: 44th Hawaii International Conference on System Sciences (2011)

161. Scheckenbach, R., Zeier, A.: Collaborative SCM in Branchen: B2B Integrationsstrategien und Realisation. Galileo, Bonn (2002)

162. Schwarz, C., Borovskiy,V., Zeier, A.: Optimizing operation scheduling for in-memory databases. In: The 2011 International Conference on Modeling, Simulation and Visualization Methods (2011)

163. Stonebraker, M.: The case for shared-nothing. IEEE Database Eng. Bull. **9**, 4–9 (1986)

164. Stonebraker, M., Abadi, D.J., Batkin, A., Chen, X., Cherniack, M., Ferreira, M., Lau, E., Lin, A., Madden, S., O'Neil, E.J., O'Neil, P.E., Rasin, A., Tran, N., Zdonik, S.B.: C-Store: a column-oriented DBMS. In: VLDB, pp. 553–564 (2005)

165. Stonebraker, M., Madden, S., Abadi, D., Harizopoulos, S., Hachem, N., Helland, P.: The End of an Architectural Era (It's Time for a Complete Rewrite). In: VLDB, pp. 1150–1160 (2007)

166. Stonebreaker, M., Rowe, L., Hirohama, M.: The implementation of postgres. IEEE Trans. Knowl. Data Eng. **2**, 125–142 (1990)

167. Sundell, H., Tsigas, P.: Fast and lock-free concurrent priority queues for multi-thread systems. In: IPDPS, p. 84 (2003)

168. Sutter, H.: The free lunch is over: a fundamental turn toward concurrency in software. Dr. Dobb's J. **30** (2005)

169. Taniar, D., Leung, C., Rahayu, J., Goel, S.: High Performance Parallel Database Processing and Grid Databases. Wiley, New York (2008)

170. Thomsen, E.: OLAP Solutions: Building Multidimensional Information. Wiley, New York (2002)

171. Thomson, A., Abadi, D.J.: The case for determinism in database systems. PVLDB **3**, 70–80 (2010)

172. Thusoo, A., Shao, Z., Anthony, S., Borthakur, D., Jain, N., Sarma, J.S., Murthy, R., Liu, H.: Data warehousing and analytics infrastructure at facebook. In: SIGMOD Conference, pp. 1013–1020 (2010)

173. Tilera: TILEPro36 Processor. URL http://www.tilera.com/products/processors/TILEPRO36 (2008). Retrieved 14 Jan 2011

174. TPC: TPC Benchmark W (Web Commerce). Standard Specification, Version 1.8, February 19, 2002. (2002)

175. TPC: TPC Benchmark C, Standard Specification Revision 5.9. (2007)

176. TPC: TPC Benchmark E, Standard Specification Version 1.5.1 (2008)

177. TPC: TPC Benchmark H (Decision Support), Standard Specification Revision 2.7.0 (2008)

178. TPC: TPC-Energy Specification. Standard Specification Version 1.1.1 (2010)

179. Transier, F.: Algorithms and data structures for in-memory text search engines. Ph.D. thesis, University of Karlsruhe (2010)

180. Transier, F., Sanders, P.: Compressed inverted indexes for in-memory search engines. In: ALENEX (2008)

181. Vassiliadis, P., Simitsis, A., Skiadopoulos, S.: Conceptual modeling for ETL processes. In: DOLAP, pp. 14–21 (2002)

182. Vogels,W.: A head in the cloud—the power of infrastructure as a service. In: CCA (2008)

183. Walsh, K.R.: Analyzing the application ASP concept: technologies, economies, and strategies. Commun. ACM **46**, 103–107 (2003)

184. Westmann, T., Kossmann, D., Helmer, S., Moerkotte, G.: The implementation and performance of compressed databases. SIGMOD Rec. **29**, 55–67 (2000)

185. Willhalm, T., Popovici, N., Boshmaf, Y., Plattner, H., Zeier, A., Schaffner, J.: Ultra fast in-memory table scan using on-chip vector processing units. In: VLDB, pp. 385–394 (2009)

186. Winter, R.: Why Are Data Warehouses Growing So Fast? An Update on the Drivers of Data Warehouse Growth. URL http://www.b-eye-network.com/view/7188 (2008). Retrieved 14 Jan 2011

187. Wust, J., Krüger, J., Blessing, S.,Tosun, C., Zeier, A., Plattner, H.: xsellerate: Supporting sales representatives with real-time information in customer dialogs. In:IMDM2011, Proceedings zur Tagung Innovative Unternehmensanwendungen mit In-Memory Data Management, Mainz (2011)

188. Zeier, A.: Ausdifferenzierung von Supply-Chain-Management-Standardsoftware in Richtung auf Betriebstypen und Branchen unter besonderer Beruecksichtigung des SAP APO. Roell (2002)

189. Zicari, R.V.: Hadoop for Business: Interview with Mike Olson, Chief Executive Officer at Cloudera. URL http://www.odbms.org/blog/2011/04/hadoop-for-businessinterview-with-mike-olson-chief-executive-o_cer-at-cloudera/. Retrieved 2 Nov 2011

190. Zukowski, M., Heman, S., Nes, N., Boncz, P.: Super-scalar RAM-CPU cache compression. In: ICDE, p. 59 (2006)

Index

H. Plattner and A. Zeier, *In-Memory Data Management*,
DOI: 10.1007/978-3-642-29575-1, © Springer-Verlag Berlin Heidelberg 2012

Printing: Ten Brink, Meppel, The Netherlands
Binding: Stürtz, Würzburg, Germany